*Acclaim for* ROBIN MARANTZ HENIG's

## A Da

"A book both fascinating an[d] ... [becaus]e the subject is compelling and H[enig] ... [thrillin]g because it puts to rest any no[tion] ... their environment."
—*Boston Globe*

"Thorough, balanced. . . . Henig makes our relentless vulnerability to emerging viruses clear . . . [with] a wealth of information and . . . even the occasional thrill."
—*Washington Post*

"Electric with true anecdotes that will thrill any [science fiction] enthusiast and alarm anyone with any imagination, this narrative is an invaluable window on the complex issues of global health."
—*Booklist*

"A racy voyage along . . . viral frontiers."
—*Journal of the American Medical Association*

"A wonderfully well written book that makes the fascination and excitement of microbiology and infectious disease accessible to a broad range of interested readers. . . . An exciting book . . . and very approachable source of up–to–date information on some of the most pressing health issues of our time."
—June E. Osborn, M.D.,University of Michigan
School of Public Health

"Complex science made not just accessible but wholly fascinating, and communicated with a sense of urgency yet without sensationalism."
—*Kirkus Reviews*

"A vital and stimulating book. Once picked up it is difficult to put down—for my money, more gripping than *Jurassic Park*."
—*Nature*

"*A Dancing Matrix* captures the excitement of the newly recognized problem of emerging viral infections. The author's writing captures the suspense of some of the most important current medical dramas. Highly recommended."
—Bernard N. Fields, M.D.,
Harvard Medical School

ROBIN MARANTZ HENIG

# A Dancing Matrix

Robin Marantz Henig, a medical writer and the author of four previous books, is a frequent contributor to *The New York Times Magazine*, the *Washington Post*, *Mirabella*, and *Self*. For *A Dancing Matrix* she received the June Roth Memorial Award and the 1994 Author of the Year Award—both from the American Society of Journalists and Authors. A graduate of Cornell who earned a master's degree in journalism from Northwestern, she lives in Takoma Park, Maryland, with her husband and two daughters. She is currently at work on a history of the Harvard School of Public Health.

# A
# DANCING
# MATRIX

*How Science Confronts Emerging Viruses*

*by*

# ROBIN MARANTZ HENIG

VINTAGE BOOKS
*A Division of Random House, Inc.*
*New York*

*For Jeff*
*and for our sweet girls,*
*Jessica and Samantha*

All rights reserved under International and Pan-American Copyright Conventions.
Published in the United States by Vintage Books, a division of Random House, Inc.,
New York, and simultaneously in Canada by Random House of Canada Limited,
Toronto. Originally published in hardcover, in slightly different form, by
Alfred A. Knopf, Inc., New York, in 1993.

Library of Congress Cataloging-in-Publication Data
Henig, Robin Marantz.
A dancing matrix: how science confronts emerging viruses / by
Robin Marantz Henig.
p.    cm.
Previously published: New York: Knopf, 1993 with subtitle:
Voyages along the viral frontier.
Includes bibliographical references and index.
ISBN 0-679-73083-4
1. Virology—Popular works.  2. Medical virology—Popular works.
I. Title.
QR364.H46   1994
576'.64—dc20      93-27553
CIP

*Book design by Virginia Tan*

Manufactured in the United States of America

# Contents

We live in a dancing matrix of viruses; they dart, rather like bees, from organism to organism, from plant to insect to mammal to me and back again . . . passing around heredity as though at a great party. They may be a mechanism for keeping new, mutant kinds of DNA in the widest circulation among us. If this is true, the odd virus disease, on which we must focus so much of our attention in medicine, may be looked on as an accident, something dropped.

—Lewis Thomas, M.D.,
*The Lives of a Cell*

# *Introduction*

The first weird death was in a 21-year-old man in New Mexico. He had been perfectly healthy until a few days earlier, when he started experiencing symptoms of a mild flu: malaise, achiness, low-grade fever. While driving with his family on a morning in May 1993, the young man started to have trouble breathing; by the time his parents pulled off the highway and into a convenience store to call an ambulance, he was panting, unable to get enough air into his lungs. He stopped breathing on the way to the hospital, and couldn't be revived.

The emergency room staff contacted the local health authorities because this young man's death was eerily familiar to one that had taken place in the same ER just one month earlier: a healthy young person (in that case a woman in her 30s) suffers from a mild case of flu, but before recovering suddenly has trouble breathing and dies. When the local deputy medical examiner, Richard Malone, talked with the young man's family about the coincidence, he was spooked by an ever stranger one. The young man had been on his way to his fiancée's funeral. She had died the week before in exactly the same way.

Malone knew he had a mystery on his hands. He called the funeral home and asked that the fiancée's burial be delayed; by that night the bodies of the two young people were on a plane to Albuquerque to be autopsied. In the month to come, nine more young people died. Beyond their initial flu-like symptoms and the rapidity of their decline, these early victims had something else in common: They all lived in New Mexico or Arizona, and most of them were Navajos.

Clearly, this was no ordinary flu. The outbreak occurred in late spring, months after the end of flu season. And it was proving far more deadly than influenza, leading to fluid buildup in the lungs, respiratory failure, and, in more than half the cases, death. Most puzzling, the people dying were the very people who usually take influenza in stride—the young and the healthy. One victim, a thirteen-year-old girl, collapsed while dancing at a graduation party; she died the next day. A young adult died in the middle of buying groceries.

The outbreak occurred in the spring of 1993, in a region of the southwestern United States known as Four Corners, because it's where four states—Arizona, New Mexico, Utah, and Colorado—intersect. Spread across Four Corners is the sprawling twenty-six-thousand-square-mile Navajo Reservation. By early June, eighteen people—most of them in their teens and twenties—had gotten sick. And of the eleven who died in that first frightening wave, seven were Navajo.

Tribal elders believed they had an explanation for the mysterious plague. They had noticed that, for the first time in memory, the piñon trees on the reservation had been in full flower for the entire year. Rodents that eat the sweet meat of the piñon nut had become fat and plentiful, their population explosion boosted not only by a rich food supply but also by a rainy spring that kept the ground soft and good for burrowing. The Navajos' explanation was soon borne out. The rodents—especially a species known as the deer mouse (*Permomyscus maniculatus*)—were harboring a deadly strain of Hantavirus, a type of virus previously thought to be confined to Asia and to port cities in the United States. Anyone living in close proximity with the deer mouse, or stirring up dry dirt particles that had previously been soaked in rodent urine contaminated by the virus, was likely to breathe in enough Hantavirus to get sick.

The virus's eruption that dreadful spring offers the most recent bit of evidence of something scientists had already started to discern: the delicate balance among human beings, the global environment, and our most challenging foe, the virus. It is the latest example of a group of destructive microbes that have come to be known as "emerging viruses."

An emerging virus, of which the most notorious is the one that causes AIDS, is a virus that crosses species or geographic boundaries and shows up with unprecedented virulence in unexpected places. Like the southwestern Hantavirus, most emerging viruses have been percolating beneath the surface for quite some time before they come to our attention. The change from harmless virus to emerging, and threatening, virus could have one of several causes. It may be due to an actual mutation in one of the virus's genes, which changes its ability to infect—though this scenario happens quite rarely. Alternatively, it may result from changes in the natural environment, including patterns of rainfall and temperature, which interfere with the balance between predators and their prey and give an unchanged virus a new niche in which to thrive. This is what seems to have happened in the Four Corners outbreak, with the flowering of the piñon trees and the consequent deer mouse baby boom.

Or, as occurs most often, a new virus may emerge because of the things people do—building roads into the rain forest, transporting microbes from one region to another, erecting housing on the edge of woodland—that change our relationship to the environment and expose new populations of people to viruses they'd never seen before. This last cause, the intrusion of humankind into the natural order of things, seems now to be the single most important factor in the emergence of new viruses. The good news is that this factor is within our collective power to change.

Every decision made anywhere in any sphere of life—environmental, political, demographic, economic, military—carries with it ramifications about disease that reverberate around the world. When the Aswân Dam was built in Egypt, the new body of still water allowed mosquitoes to thrive, and the viruses they carried became a new threat. When used tires were shipped from Japan to Texas, mosquitoes hitched a ride in the wet rims to a new continent; because the mosquitoes can carry viruses never before seen by Texans, their presence was also a new threat to public health. And when the city borders of Seoul were pushed further into the countryside, urban Koreans were exposed to a virus that field mice had been carrying for centuries—and many contracted a raging hemorrhagic fever that kills at least ten percent of its victims.

In the 1960s and 1970s, neither public health officials nor scientists realized that "progress" could have such consequences. With a hubris worthy of the best Greek tragedy, we Westerners thought we were nearly invincible, at least as far as infectious disease was concerned. We thought we could interfere with ecosystems, and ship goods and people around the world, with little regard for the effect not only on the balance of nature, but on our own health.

Scientists at the time tended to see infectious diseases as a series of problems to solve, ticking off victories like notches on a gun belt. Through science, we had conquered polio, eliminated smallpox, created vaccines for most childhood diseases, and devised "miracle drugs"—antibiotics—for whatever infections managed to slip through the vaccine safety net. "The war against infectious diseases has been won," declared the U.S. surgeon general, William H. Stewart, in 1969. And Sir Macfarlane Burnet, the Australian virologist who shared a Nobel Prize for his hypothesis of acquired immunological tolerance—which paved the way for organ transplantation—offered a similar notion in the introduction to his textbook's third edition. Remarking on how much had changed since the original version of *The Natural History of Infectious Disease* was published more than twenty years earlier, Burnet wrote in 1962 that the late twentieth century would be witness to "the virtual elimination of infectious disease as a significant factor in social life." To write about infectious disease, he said, "is almost to write of something that has passed into history."

AIDS offered the most obvious proof that victory over infectious disease had been declared prematurely. The minuscule human immunodeficiency virus, made of little more than a few strands of pure genetic material, showed us just how mortal we are. It showed us how difficult it is to anticipate new pathogens before they already have us in their grip. And it showed us—in one more frontal assault on our collective hubris—that in the natural order of things, we human beings seem to be quite tangential.

The AIDS virus has managed to persist and thrive, multiplying today in an estimated ten million people around the world, despite the fact that it eventually kills a huge proportion of its human hosts. Its lethality is of no consequence to the virus, as long as it infects a new host before its old host dies—something that is all but assured by the virus's long latent period. Viewed this way, as a host species that helps in the perpetuation of a parasitic organism, human beings begin to look like little more than, as a British biologist puts it, "colonies of viruses."

The surprising role of viruses in nature is one central theme of this book. From the perspective of a virus, a host—even so lofty a host as a human being—is simply a convenient way to make more viruses. When humans get infected, it's not because the virus is especially "wily" or "devious," adjectives traditionally applied to this complicated pathogen. Humans get infected, quite often, simply because they have put themselves in some virus's way.

Richard Krause has been preaching this gospel for more than two decades. And preaching is the apt term; with his Ohio twang and slicked-back white hair, Krause looks and sounds like a midwestern evangelist. But for many years, even though he spoke from positions of authority—including, for eight of those years, as director of the National Institute of Allergy and Infectious Diseases (NIAID), a branch of the U.S. National Institutes of Health (NIH)—Krause found his audiences either skeptical or uninterested. It took the tragedy of AIDS to get folks to listen.

"A major dislocation in the social structure—love, hate, peace, war, urbanization, overpopulation, economic depression, people having so much leisure they sleep with five different people a night—whatever it is that puts a stress on the ecological system, can alter the equilibrium between man and microbes," Krause has been saying. "Such great dislocations can lead to plagues and epidemics, more often than not caused by microbes that already reside right on our doorstep."

As we examine how and why new viruses emerge, Krause (pronounced *krau*-see) will be one of our guides. Now a senior scientist at the Fogarty International Center at NIH, Krause is often called upon to advise scientists and administrators at NIAID, which has become the U.S. government's lead agency for AIDS research. His slim volume of essays published in 1981, *The Restless Tide: The Persistent Challenge of the Microbial World*, is still cited by the nation's leading infectious disease experts, among them the current head of NIAID, Anthony Fauci. In fact, when Fauci had his staff prepare posters to illustrate his annual budget appeal to Congress in 1990, they used Krause's little book for inspiration. They drew an illustration called "The Restless Tide of Diseases," a series of infection-speckled tidal waves with a big question mark under the wave labeled "21st Century." Fauci later framed the poster and presented it to his predecessor as a personal memento.

If Krause, in his late sixties, is the white-haired sage in the field of emerging viruses, then Stephen Morse is its young activist. Morse is an assistant professor of virology at The Rockefeller University in

New York (where Krause, coincidentally, made his reputation as a bacteriologist in the 1950s and 1960s). He looks younger than his early forties, probably because of his earnest manner, big horn-rimmed glasses, and high-pitched voice. You can imagine that he looked and acted much the same when he was a precocious student at New York's Bronx High School of Science, and when he graduated, at the age of nineteen, from the city's premier public university, City College. Now he spends most of his time thinking, writing, and talking about emerging viruses. (He believes he even coined the term, but in a characteristic mixture of modesty and precision he hesitates to state it that baldly.) Morse coordinated a conference on the subject in 1989, chaired the viral working group for a committee on emerging microbes sponsored by the National Academy of Sciences, has edited a scholarly book on viral emergence and one on viral evolution, and manages (though just barely) to squeeze in laboratory time to work on his "real" research on a mouse herpesvirus.

Though a generation divides Dick Krause and Steve Morse, they are bound by a common intellectual drive to make sense of infectious diseases, to anticipate their evolution, to figure out where human beings fit into the scheme of things. Viruses can be an important Rosetta stone in the task of deciphering the hieroglyphs of genetics, ecological relationships, and human behavior, all puzzles that must be untangled before the larger, more looming questions can be addressed.

New viruses have always emerged, since the beginning of recorded time. But the pace might well be quickening as we continue to encroach on nature in ever-more insidious ways. Human actions from forest clearing to genetic recombination are making viral catastrophes increasingly more likely; the result could well be the ecological equivalent of a nuclear holocaust.

It certainly seems as though emerging viruses, and the corresponding academic interest in studying and anticipating them, have gained new attention in the single year since this book was originally published in its hardcover edition. There was, first of all, the southwest Hantavirus outbreak, as well as a mysterious illness in Cuba that looked for a while as though it was due to a new tropical virus. (It turned out that it wasn't.) As the instances of new viruses became apparent—a projected increase in arbovirus disease after the Midwest flood in the summer of 1993, an unseasonable outbreak of a particularly virulent strain of Beijing flu originating in Louisiana, four or five months before such outbreaks usually begin—the scientific community started to award this field of study the attention it deserves.

While the hardcover edition of this book was the first exploration of emerging viruses for a lay readership, its publication coincided with two scholarly books on the same topic: an updated compilation of papers presented at Steve Morse's 1989 conference; and the final report of the committee on emerging microbes convened by the National Academy of Sciences. Since the appearance of these three books, scientists have moved beyond talking and writing about emerging viruses, and are actually starting to do something about them. In 1993, two research projects were instituted—one in Papua New Guinea, the other in Brazil—to track the course of emerging viruses before they get out of control.

New logging operations in Papua New Guinea, which are cutting swaths through the lush rain forest that had until recently covered nearly three-quarters of the island, are the object of scrutiny in the first such project, funded by the MacArthur Foundation. Its goal is to accumulate data over the next three to six years to test the theory that new viruses emerge when environmental conditions change—a proposition that will be repeated again and again in this book, but one for which there is to date only circumstantial evidence. Field workers will assess the health of one thousand people living in four villages, two in a region slated for clearing, two in the virgin forest. They will look for evidence of exposure to the AIDS virus, herpesvirus, Ross River virus, and dengue fever virus, as well as the nonviral microbes that cause typhoid, malaria, and sexually-transmitted diseases. They will also look for viruses in forest-dwelling animals—mosquitoes, birds, rodents, and bats. These assessments will be repeated periodically to see if anything new turns up as the forest ecology changes.

The Brazilian project will involve essentially the same prospective investigation, but in the Amazon rain forest. Supported by the Rockefeller Foundation and directed by Robert Shope, the Yale virologist we will meet later in this book, scientists will study three hundred residents of Abaetetuba, a municipality just inland from the coastal city of Belem. Significantly, this project will include an experimental component as well, dividing the rain forest into four quadrants. In the first quadrant, residents will do what they have always done: clear the forest and plant cash crops. In the second, they will clear the forest, plant cash crops, and also plant fruit trees native to the rain forest. In the third, no crops will be planted, though a few rain forest trees will be cleared to make way for native fruit trees. And in the fourth, the forest will remain untouched. The goal, according to Shope, is to discover the least intrusive way—in terms of public health—to settle Amazonia.

"The single biggest threat to man's continued dominance on the planet is the virus," writes Joshua Lederberg, bacteriologist, Nobel laureate, and president emeritus at Rockefeller. Lederberg co-chaired, with Robert Shope, the National Academy of Sciences committee on emerging microbes. And he is fond of striking down our collective hubris with his warnings about the magnitude of the viral threat. "The survival of humanity," Lederberg notes often in his speeches and his writings, "is not preordained."

But neither is humanity doomed. A historical perspective on new disease might give us cause for concern, but it offers at least some cause for optimism, too. In the first place, just as viruses can emerge over the course of evolution, so can they disappear. The English sweating sickness, for instance, spread wildly in the fifteenth and sixteenth centuries, but the virus presumed to have caused it now seems to be extinct. In the second place, just as human manipulation can be blamed for creating these new threats to our health, so can it be praised for actually eliminating certain viral diseases. The biggest virus victory was against smallpox, which was eradicated in 1977, after a worldwide vaccination program. In the third place, scientists are working to set in motion systems of surveillance that they hope will mean they won't be caught by surprise again. "Although virologists cannot yet predict specific disease outbreaks," says Morse, "we now understand many of the factors leading to them. And because we have a better grasp of their origins, for the first time we are in a position to do something about emerging diseases at fairly early stages."

The current scientific understanding of the patterns of emerging viruses—generated, in large measure, in the wake of AIDS—probably helped speed the response to the recent Hantavirus outbreak. After the first clusters of patients were spotted in New Mexico, it was only a matter of weeks before the new virus, and the rodent that harbored it, were identified. This is in stark contrast to the comparatively slug-

*Introduction*

gish response to AIDS, with nearly three years passing between the
first reported deaths and identification of the human immunodeficiency virus.

In the pages to come, we will meet some of the men and women
from around the world who are working to stay ahead of these new
pathogens. They are laboratory scientists, fieldworkers, epidemiologists, veterinarians, and physicians, and for the first time they're trying
to develop a common vocabulary with which to mitigate a common
foe. The study of emerging viruses might, in fact, move biologists
away from their reverence for molecular biology, with its increasingly
reductionistic view of the gene as an explanation for everything, toward a more generalized, more multidisciplinary biology. This new
biology would be one that integrates the study of genetics with the
study of organisms as a whole—how they work, interact, coexist, and
coevolve.

The viral frontier is a place of both dangers and opportunities.
Sometimes the scientists involved can get cocky or overconfident;
sometimes they can get overwhelmed with how little they really know.
Krause has been saying for years something that his juniors are just
recognizing as true: "We cannot count on the medical solutions of the
past to solve the problems of the future. Indeed, the re-emergence of
old problems in new garments is programmed in the genetic machinery
of evolution."

In this book I will try to show scientists as people who have
triumphs and who make mistakes, and to show the science of "viral
emergence" as one that is itself just starting to emerge. As I do so, I
will try to avoid the hubris that has plagued similar pursuits in the
past, remembering that as far as the virus is concerned, we human
beings—teachers or tradesmen, plumbers or patriarchs, molecular biologists or journalists—are little more than a handy way to reproduce.

# PART ONE

# THE TINIEST MENACE

# I

# *Why New Viruses Emerge*

On February 3, 1989, a forty-three-year-old mechanical engineer living in a Chicago suburb went to his health maintenance organization complaining of fever, chills, sore throat, and muscle aches. It was winter in the midwest, and flu was going around; the engineer went home with a diagnosis of influenza. Within a week he was back at the clinic. A different doctor thought it might be strep throat, so he prescribed penicillin. A few days later, with a throat so sore he could barely swallow, a raging fever, and symptoms that now included bloody diarrhea, the engineer went to the clinic again. This time he was seen by a specialist in digestive disorders who was completing his studies at a prestigious Chicago medical school and moonlighting at the HMO to make some extra money. His diagnosis was a bit more creative: influenza (to explain the fever and sore throat) with hemorrhoids (to explain the bleeding).

These doctors were all doing what they had been taught to do in medical school. When physicians-in-training learn the art of differential diagnosis—making the right diagnosis from among the many that are possible for any set of symptoms—they are offered a handy aphorism. "When you hear hoofbeats in Central Park," the saying goes, "don't expect zebras." The slogan is meant to provide medical students with a common-sense approach to diagnosis. Expect the expected, it says; don't be so blinded by the obscure that you forget about the ordinary. But as the world gets smaller, such advice might become outdated. Doctors can run into trouble expecting horses when zebras start arriving at the doorstep.

Two days after the flu-with-hemorrhoids diagnosis, the engineer

3

was admitted to Central Du Page Hospital in Winfield, Illinois. His fever was 103°F, his pulse rate was four times normal, his body was in shock. By the following morning, he was dead.

The fact that the man was Nigerian by birth and had a thick accent had never tipped off his first three doctors to a connection to Africa, the continent where many unfamiliar diseases originate. No one had asked the man whether he had traveled out of the country recently. But that turned out to be the crucial question. The engineer had gone home to Nigeria on January 18 to attend his mother's funeral. He didn't know it at the time, but what had killed his mother was an unusual and highly virulent African virus. When he flew back to the United States on January 31, arriving at O'Hare International Airport on the morning of February 1, the engineer felt fine. But he was already infected with the deadly virus. It was just a matter of time before the fevers and the hemorrhaging began.

Only when he was hospitalized did he finally get a correct diagnosis. Robert Chase, an infectious disease specialist who consulted on the case, thought to ask the man's wife about his travel history. He then placed a worried phone call to Joseph McCormick at the Centers for Disease Control (CDC) in Atlanta, who had studied viral hemorrhagic fevers in Africa in the late 1970s. "Is there a disease in western Africa that causes high fever, sore throat, liver damage, and internal hemorrhaging?" the physician asked. McCormick's answer: "You bet there is."

It turned out that the engineer had the same disease that had probably killed his mother: Lassa fever. During the time the man had been home for the funeral, his father and several cousins also died of Lassa fever, and two sisters got sick and recovered. The man's brother, a physician in Nigeria, died of Lassa soon after the engineer had gone back to America. The engineer's death was the first signal to the rest of the world that Nigeria was in the middle of an epidemic.

First recorded in Nigeria in 1969, Lassa fever is caused by exposure to a rat that, when infected, secretes the Lassa virus in its saliva and urine. It is easily spread from rat to rat, and from rat to human; the virus is transported through breaks in the skin, so people living in homes where rats get into the food supply or urinate near the house are likely to be exposed to lots of infectious Lassa virus. Such conditions are common in many poor communities in Africa, especially since the rat that carries Lassa fever, known as the "soft-furred" rat, is one of the few wild rodents known that is perfectly comfortable living around people. But easy as it is to pass among rats and from rats to humans, Lassa fever is quite difficult to pass among people without a rodent intermediary, except for the most intimate contact. Over the past twenty years, in sporadic flare-ups scattered across Africa, Lassa fever has killed nearly 40 percent of the severest cases. In 1989 alone, it was responsible for five thousand deaths in western Africa.

Within hours after Chase's phone call to CDC, the engineer was put on a respirator to ease his breathing. Full doses of vasopressors (blood vessel constrictors) were pumped into him to fight the hemorrhaging. An emergency supply of experimental antiviral medications was ordered from the manufacturer and was being flown to his bedside. But none of the muscle of modern medicine could combat this primitive virus. "He died on a respirator, with a Swan-Ganz catheter in him [a sterile tube inserted in the pulmonary artery to monitor impending heart failure], in a major-league hospital, with a team of all sorts of experts standing around," says McCormick. "And they couldn't save him. They told me, 'He didn't respond to anything we gave him.'"

What if that first doctor had made the correct diagnosis? Wouldn't earlier treatment have saved his life? "I'm not convinced of that," McCormick says. "We don't have any experience with antivirals in Nigeria, and we have reason to believe that Lassa fever is a little more virulent in Nigeria," where the engineer got infected, than in other countries in which it has been studied, such as Sierra Leone.

On February 16, the day the engineer died, McCormick flew to the man's hometown to stem the rising panic. The man had been an active church member in the community of Glen Ellyn, and many friends had visited him while he was sick. The hospital workers who cared for him hadn't known to use special precautions in handling his bedpans and needles. There was the terrible dread that Lassa fever would start to spread in Illinois.

McCormick prescribed ribavirin, the experimental drug that had arrived too late for the engineer, to his widow and their six young children. The mother and the children, aged seven to seventeen, had all shared the man's plate, changed his linens, and stayed at his bedside during his two-week illness. But either because of the ribavirin, or because Lassa fever is difficult to transmit without rats in the environment, none of them developed Lassa fever.

5

As far as the CDC is concerned, the story essentially ends there. No one other than the engineer died; no one else even got sick. (One of his nurses developed flu symptoms on February 17, which gave her quite a scare, but she recovered within forty-eight hours.) But in bare outline, this is precisely the kind of situation many virologists are worried about. In such a scenario, a new, potentially quite lethal virus makes its way across great geographic distances to begin threatening a previously unexposed population. If Lassa fever in this country proved as easily transmitted as it seems to be in Nigeria, this single case alone could have set off the next AIDS.

An emerging virus isn't necessarily a new virus; it's just new to the community that is threatened by it. When a virus mutates spontaneously, or crosses species or geographic borders, it imperils a population that was never before exposed. And it's in the first wave of an emerging virus epidemic that the most damage is done—as occurred with the rabbit virus, deliberately introduced into Australia in 1950 to wipe out an imported rabbit population that had grown totally out of control. A British gentleman had first imported a few dozen rabbits from England "for sport" in the 1850s, but within twenty years they had reproduced so wildly—the rabbit has no natural predators in Australia—that it was a threat to crops, pastures, and the survival of less prolific species. So in 1950, a highly virulent strain of a rabbit virus known as myxomatosis was brought to Australia from Brazil, where it was a relatively harmless tumor-causing agent.

For a while, it looked as if myxomatosis would do the trick. In the first weeks after the virus was released, myxomatosis killed Australian rabbits with a fatality rate of 99.8 percent. But just one year later, the fatality rate was down to 90 percent; by 1958, it was a mere 25 percent. Rapid evolutionary changes in both the myxomatosis virus and the Australian rabbits had led to adaptations that allowed pathogen and host to coexist in relative equilibrium. "We now see a fascinating interplay between genetic changes in host and virus," wrote Frank Fenner, a prominent Australian virologist, more than thirty years after the myxomatosis introduction, which he had helped coordinate. "It is to be hoped that observations can be continued, at intervals of perhaps a decade, to examine this unique model of the continuing evolution of an infectious disease."

6

Myxomatosis qualifies as an emerging virus disease for Australian rabbits. But it did not just generate spontaneously. Scientists found the virus in a laboratory in Brazil, where it was killing rabbits imported from Europe for experimentation but sparing native rabbits in the surrounding woods near São Paulo. They isolated the virus from the rabbits that died, grew it in the laboratory, and brought it into Australia to inject it into rabbits living in warrens near the upper Murray River. On river flats twenty miles away, rabbits began dying of myxomatosis—not because of the inoculations alone, but because an important vector of the virus, the mosquito, also existed in the region. (A vector is an insect or animal that can bite a viral victim, take in some of the virus, and allow the virus to grow in its own body and be passed on later through another bite.) So human intervention was necessary, but not sufficient, for touching off the Australian plague of myxomatosis. The introduction of the virus was stunning in its early success—indeed, in three short months the myxomatosis plague had spread across a geographic area the size of western Europe—but human actions also required one critical natural condition: the prior existence of the vector mosquito.

The pattern of the myxomatosis plague is probably followed in other emerging virus diseases, animal and human alike. When a new population is hit, the immediate effect is almost total devastation. Within a few years, though, the animal host evolves to become collectively less vulnerable. Those not killed in the first wave either are genetically resistant or have become resistant after developing antibodies against an initial infection. And while the animal host evolves, the virus is evolving too. The most lethal strains kill their hosts so rapidly that they themselves die out; they are unable to infect a second host before the first host dies. This gives an advantage to strains of virus that kill more slowly. After the host animal has gone through about six generations, this coevolution usually works its way to a state approaching equilibrium: one with a virus that is less virulent, so its host stays healthy enough to pass it around, and with a host that is less vulnerable, so it survives despite this native infection. To reach this equilibrium usually takes about six generations—no time at all for fast-breeding rabbits, which become parents by the age of six to ten months, but at least 125 years for human beings.

Similarly, emerging viruses that infect humans have generally been in nature, percolating beneath the surface, for quite some time. Some-

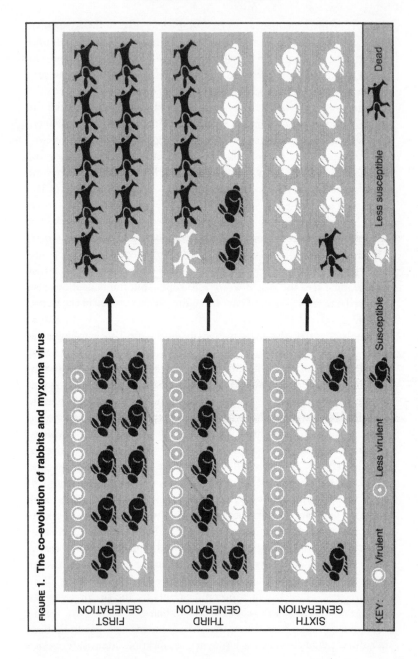

FIGURE 1. The co-evolution of rabbits and myxoma virus

FIRST GENERATION

THIRD GENERATION

SIXTH GENERATION

KEY: Virulent · Less virulent · Susceptible · Less susceptible · Dead

thing about the virus then changes: its actual genetic configuration, its position in the ecological balance, or its proximity to a species not previously encountered. When a virus "emerges," it slides out of its original niche and begins infecting a new population. This usually occurs as a result of an actual mutation in one of the virus's genes, which changes its ability to infect; of changes in the environment, including patterns of rainfall and temperature, which interfere with the balance between certain predators and their prey and give an unchanged virus a new niche in which to thrive; or of the things people do—building roads into the rain forest, transporting microbes from one region to another, erecting housing on the edge of woodland—that change man's relationship to the environment and expose new populations of people to viruses they'd never seen before.

These last two mechanisms—in which either man or nature shifts sufficiently to change the balance between the two—are not inherently as interesting as the first. It seems we would much rather imagine weird new microbes arising out of the blue than think about the more mundane matter of how our actions affect our health. From the best-selling *Andromeda Strain*—Michael Crichton's 1969 novel about a new infectious agent falling from outer space—to inflamed political rhetoric about a new microbe getting loose from a gene-splicing laboratory, mutations have taken center stage in our collective nightmare. But mutations are almost never responsible for emerging viruses, which makes one wonder why we are so captivated by the image of man as hapless victim of nature, rather than the other way around. The fact is, as far as nature is concerned, human beings are not at the center of things, and nowhere is this more in evidence than in the study of new viruses. Occasionally, we bumble into some virus's way, and we encounter catastrophe. But that is not what the virus was doing there originally. We created the mishap, and our suffering is of no consequence to the virus itself. As the essayist Lewis Thomas has put it, "We have always been a relatively minor interest of the vast microbial world. . . . Disease usually results from inconclusive negotiations for symbiosis, an overstepping of the line by one side or the other, a biologic misinterpretation of borders."

Figure 1. When myxoma virus was introduced into Australia in 1950, it killed 99.8 percent of exposed rabbits during the first year. Those few rabbits that survived (1st generation) were genetically resistant. When these rabbits reproduced (3rd generation), the proportion of resistant rabbits in the population increased. At the same time, the most virulent viruses died out as their hosts died, leading to a greater proportion of less virulent myxoma virus (3rd generation). The result of this co-evolution was that the mortality rate of myxomatosis steadily declined, stabilizing at about 25 percent (6th generation).

Viruses and their natural hosts—which are almost always *not* human beings—could go on quite happily without our interference. But interfere we do, over and over again. And it's when this happens that we get ourselves in trouble, as we have been doing for at least five hundred years, ever since Columbus—and probably for many hundreds of years before that. The causes of the great diseases of antiquity still remain mysteries to modern epidemiologists, but the health effects of the discovery of the New World have been clearly documented. When European explorers first came to the Americas, they brought along in their bodies and their baggage deadly viruses that the Native Americans had never before seen. The epidemics that resulted helped shape the course of history.

Cortés's small band of warriors, for instance, probably would never have been able to conquer Mexico in 1521 had it not been for smallpox. The disease killed more than one-third of the Aztecs that year; the tribespeople considered such massive devastation to be the gods' punishment for some unintentional transgression. This was confirmed, in the Aztecs' mind, by the bizarre pattern of death in the wake of this hideous pestilence: none of the Spaniards died of smallpox. There was of course a simple biological explanation for this—the Spaniards had all been exposed to the virus in childhood, so they were immune—but the Aztecs thought it was a kind of holy magic. They were dying, they believed, because they deserved to die, and the white men were living because they somehow had earned favor with the gods. As William McNeill put it in his classic historical work *Plagues and Peoples*, "From the Amerindian point of view, stunned acquiescence in Spanish superiority was the only possible response."

As ships continued to crisscross the oceans over the centuries, viruses continued to stow away on board. Sometimes, as with measles and smallpox, the virus took up residence in the sailors themselves. When the sick sailors got off at the port of call, their very presence in the new city was enough to start spreading disease.

But not every virus is passed directly from one person to another. Many of them involve an intermediate insect or animal carrier—a vector—that can harbor the virus with no ill effect and, when it bites its next victim, pass on the virus as efficiently as would a hypodermic syringe filled with viral fluid. Vector-borne diseases can never be transmitted by people breathing on each other, as measles and smallpox can. They can never be passed on by people having sex with each

other, as most of the herpesviruses can. They are contagious only
through an intermediary. So for these diseases to be introduced into
a new geographic location, two separate events must occur in close
succession: the virus must be imported (usually inside a human patient), and the vector (usually an arthropod, but occasionally a small
rodent) must be imported too, or must already be present on native
soil.

The slave ships of the seventeenth century turned out to offer the
ideal conveyance for bringing one type of virus vector, the notorious
*Aedes aegypti* mosquito, from its native West Africa to the rest of the
world. *Aedes aegypti* can carry and transmit the virus of yellow fever,
a debilitating disease that causes high fever, headache, nausea and
vomiting, and, in the worst cases, jaundice (a sign of liver failure),
proteinuria (protein in the urine, a sign of kidney failure), and so-called
black vomit (which indicates internal bleeding). The yellow fever mosquito is highly domesticated and will breed only in artificial containers;
it won't lay its eggs in water with a muddy or sandy bottom. When
ships began transporting Africans to the Caribbean, where they were
sold as slaves, the open water barrels on board were propitious breeding grounds for *Aedes aegypti* larvae. And once the boats docked in the
Caribbean, the warm climate proved quite hospitable for this tropical
Old World mosquito.

The mosquito vector was one of the two critical ingredients needed
for yellow fever to take hold. The other one was the virus itself. That
came to the Caribbean in the 1640s, when sailors on a slave ship took
ill during the long voyage from West Africa to Havana. They were
suffering from "yellow jack," so called because an infected ship was
required to fly a yellow flag, or jack, as a warning to those onshore.
As more and more European sailors died, mosquitoes continued to
breed in the water casks; new, nonimmune mosquitoes continued
to hatch, and as they bit the infected sailors, the pernicious cycle
dragged on. When the boat finally docked in Cuba, it released dozens
of sick men and bucketloads of young mosquitoes, some of which were
already viremic (carrying virus in their blood). When the viruses in
these carrier mosquitoes migrated into the salivary glands—as infectious agents do in all mosquito-borne diseases—and the mosquitoes bit
the nonimmune Cuban natives, yellow fever virus was passed directly
into the Cubans' bloodstreams. So in 1648, the hemisphere's first cases
of yellow fever were reported in the port cities of Cuba and Yucatán,

thus establishing in the New World a disease that had already ravaged the Old.

In recent years, another mosquito, known informally as the Asian tiger mosquito, seems to be replacing *Aedes aegypti* in both Latin America and the southern United States. (Its made-for-tabloids nickname comes from its coloration—white and black stripes—as well as from its big-cat tendency to attack and bite with abandon.) The tiger mosquito arrived in 1985 on a boat traveling from Japan that was carrying a shipment of used automobile tires to a dock in Texas. Stowed away on board were mosquito larvae, which are always hatched in wet areas. The rainwater that had accumulated in the tire hulls was just wet enough for the purpose. Beginning with this Texas importation, tiger mosquitoes have since spread to eighteen states, mostly in the southwestern United States. In many states, the new mosquitoes have almost wiped out the population of the native *Aedes aegypti*. That may sound comforting, but it's not: the tiger mosquito can carry the yellow fever virus too. (Not every mosquito can serve as a vector for every virus.) And in addition, the tiger mosquito, more properly called *Aedes albopictus*, can carry the virus that causes two related diseases: dengue (pronounced *deng*-ee) fever, which is relatively mild, and the more virulent dengue hemorrhagic fever.

Many experts consider dengue fever to be the most likely candidate for the next American plague. No native cases have yet arisen in the United States, but it seems only a matter of time. Now that the mosquito vector has arrived, we need only a few sick people for dengue to take hold. Once the mosquito bites an infected person, the virus passes to the mosquito's salivary glands for an important stop in its life cycle—a chance to replicate logarithmically. Then the mosquito bites an uninfected person, and the endless cycle begins.


Stephen Morse has been described as "an assistant professor straight out of Central Casting." He has that look: nearsighted, earnest, and probably incapable of running a half-mile. It's clear that Morse's life is lived largely in the mind.

He didn't set out to become an expert in emerging viruses. It all started, he says, because of a fluke. He had been at The Rockefeller University just over two years, still a research associate—after four years as an assistant professor at Rutgers—even though he was already thirty-six years old. (Postgraduate training in biomedical science goes on for a painfully long time.) He was invited to a faculty reception at the university president's house, something that the president at the time, Joshua Lederberg, liked to sponsor once a year. Standing in the living room on a winter afternoon in 1988, musing on the apartment's commanding view of the East River, Morse no doubt cut an awkward figure. If it was a day like any other day, he was wearing his standard outfit; his dress-up wardrobe is so limited that he still remembers wearing his navy blazer, blue striped shirt, and gray flannel trousers. And if it was a social occasion like any other social occasion, he was shifting his weight and looking backwards over his shoulder and generally exuding discomfort.

Then the president approached. His wife, Marguerite, had just reminded Lederberg that he had been meaning to ask an important question of the lab veterinarian. But since the vet was unable to make the party, Lederberg spontaneously decided to ask a proxy. "Should we be worried about Hantavirus?" Lederberg asked Morse. He was conveying the concern of his friend, D. Carleton Gajdusek, a scientist at the National Institute of Neurological Disorders and Stroke who had won a Nobel Prize for discovering the method of transmission for the virus-caused disorder known as kuru. Gajdusek, always aware of how bizarre viruses can be—the kuru virus was transmitted among cannibals in New Guinea who ate the brains of other kuru victims—thought Hantavirus might create a new hazard, if it came into the laboratory in wild strains of experimental rodents that usually harbor the virus in nature.

Lederberg had done his own research (for which he too had been awarded a Nobel Prize) in bacteriology. He didn't know much about viruses, especially about this strange-sounding Hantavirus. All he knew was that it primarily infected people in Asia, that it was carried in the air, and that inborn infection with Hantavirus had been reported in specially bred laboratory strains of rats in Europe that were being sold for experimentation to laboratories across the Continent. At Rockefeller, although there were no laboratories working on the virus directly, the rat that often harbored Hantavirus was the most widely used experimental species.

Morse didn't know much about the Asian virus either. But that didn't stop him from finding out. Within a few weeks he wrote a long memo to Lederberg, with scientific journal articles attached, explaining what Hantavirus was and to what extent it presented a threat to scientists experimenting on lab rodents at Rockefeller. Lederberg dashed off a four-sentence handwritten note in reply. Not only does Morse still have the note in his files, but he can quote from it verbatim: "We need some high level policy attention to what needs to be done globally to deal with the threat of emerging viruses."

With this informal pronouncement from a man he admired—and Lederberg, with a stubby white beard and gentle manner, seems to attract admiration—Morse found himself daydreaming about just how such "high level policy attention" might be focused. It was a question that concerned him too, and now that he had an indication that even Lederberg was worried, he felt an intellectual obligation to pursue the subject more seriously. "In my spare time, my thoughts would turn to the question of emerging viruses—questions like, Where did AIDS come from?" he recalls. "Whenever I was not working in the lab on mouse herpesviruses, or writing grant proposals, I would mull it over, worry it a little. It was a problem of such enormity, and unpredictability, that I was really at a loss."

Characteristically, one of Morse's first steps was to call someone who knew even less about the subject than he did, but who might offer a fresh perspective. He telephoned a Rockefeller colleague whom he barely knew—Mitchell Feigenbaum, the physicist who developed chaos theory—to see if someone from another scientific discipline might see the problem differently. "I wanted to see whether chaos could help predict viral mutation," Morse says. "I was still focused on mutation as the machine that drove viral emergence."

Chaos theory, briefly stated, looks for the pattern of predictable behavior in supposedly random events. Physical phenomena, ranging from the way water drips from a faucet to the swirling of air currents in a thunderstorm, go through a pattern of changes that have come to be called chaotic. It seemed a good hunch that chaos theory might help anticipate the apparently random events of viral mutation. Feigenbaum, though, did not think it would. Still, after their initial phone conversation he and Morse became friends, and spent many long hours over many espressos talking about the possibilities of chaos in biology. "We didn't find any answers," Morse says, "but we have had some very good times together."

Morse's interest in mutations reflected the scientific thinking of the time. By the late 1980s, scientists had become fixated on mutations as a way to explain viral threats, in much the same manner they were fixated on genes as an explanation for most of biology. Their approach proved the truth of the old dictum "Give a child a hammer, and every problem starts to look like a nail." Molecular biologists had a wonderful new hammer—recombinant DNA technology—so they began to assume that any biological problem could be accounted for by the arrangement, or misarrangement, of the genes. Since scientists could actually create mutations in the laboratory, they were inclined to believe mutations explained a great deal about viral emergence and evolution. How could they think otherwise? If mutations did not explain much, and yet mutations were precisely what they were mimicking in the lab, they might have to face the dreadful possibility that their day-to-day lab work had little bearing on what happened in real life.

"I think to some extent our focus on mutations paralyzed us," Morse says now. "Mutation is a random event, and by definition you can't predict the course of random events. A lot of fatalism crept into the field. I think basically people felt if you can't foretell the future, what's the point in trying?" Virologists settled in to studying viruses whose clinical effects were generally known, rather than trying to anticipate what new viruses might be waiting in the wings.

But for some reason, Morse still kept his eye on the periphery, wondering what would come up next. It may have been his personality: he seems by nature an optimist, a man not at all inclined to fatalism. It may have been his intelligence: here was a huge, seemingly insoluble problem that he could sink his metaphoric teeth into, bringing together his own wide-ranging interests in politics, history, geography, and philosophy. For extremely bright people, experimental biology is often intellectually unsatisfying; even when the work goes well, it involves long stretches of tedium and routine and requires a willingness to focus on small questions, with small answers. Perhaps that is why so many biologists engage in other creative pursuits. The best scientists do something else as well; they play musical instruments or contract bridge, they fool around with computers or read Victorian novels.

In some small measure, Morse chose his own intellectual "hobby" out of a kind of ego gratification. His ego, though, does not work the way a non-scientist's might; its satisfaction has little to do with wealth or fame or even tenure, none of which is best pursued through this untraditional, nonlaboratory work. Morse's ego gratification comes from something more difficult for most people to understand: the personal reward of having helped solve a puzzle of some significance to human well-being.

Morse's inspiration for how to address the problem of where new viruses come from might also seem, to the uninitiated, not so inspiring at all: he decided to host a conference. It sounds almost silly in retrospect. The image of a group of scientists talking about emerging viruses, while all around them people are dying of AIDS, is a little like picturing Nero and his violin in the middle of a Roman inferno. But conferences are the stuff of science; they are truly how scientific progress is made. Physicians and basic scientists did not even figure out that the AIDS epidemic existed until they started talking to one another at conferences in 1980 and 1981. And people musing on their own about strange new viruses—physicians, molecular biologists, ecologists, historians, epidemiologists, social scientists—did not even know they had information to share until Morse finally put them all in the same room in May 1989.

To get the conference going, Morse sought the financial support of the National Institute of Allergy and Infectious Diseases. He was lucky, he says, to find a kindred soul in John La Montagne, the institute's director for infectious diseases. Morse credits La Montagne with a visionary's ability to see the global implications of a problem even when it is so subtle that most knowledgeable observers overlook it entirely. A soft-spoken man, La Montagne at first glance seems more a conscientious bureaucrat than a visionary, writing lists on a yellow legal pad of the points he'd like to make as an interview begins. Perhaps because he started his scientific career studying the influenza virus, though, he maintains a sharp awareness of how rapidly viruses can change, and of how devastating a new epidemic can be.

Once the conference was supported, and once it transpired, the next step in creating an awareness of the issue was to get the attention of journalists, who in turn could carry the message to other professionals and to the public at large. In this regard, Morse's conference was highly successful. Reporters from *BioScience*, *Medical World News*, and *Science News*, which are read by scientists and clinicians, followed up the meeting with detailed reports about the threat of emerging viruses. The medical columnist of the *New York Times*, Lawrence Altman, was also there; he wrote a long article about the conference the following week. (The article was quite complete, except that it called

Joshua Lederberg, rather than Stephen Morse, "a chief architect of the meeting." Morse says when he read that statement, with the accompanying photo of Lederberg, he was "crushed" that there was no mention in his hometown paper of his own central role in conceptualizing, organizing, and chairing the conference.)

Within months, other magazines had picked up on the professional journals' descriptions of this new field of study. The scientifically literate layman had plenty of opportunity to read about it; *Discover, Omni, The Sciences,* and *Issues in Science and Technology* all ran feature articles about emerging viruses in 1990. (The last two were written by Morse.) In early 1991, Morse attained his fifteen minutes of fame: *Business Week* ran a piece about emerging viruses, accompanied by a flattering photo of him, which quoted from him no fewer than three times. And around the same time, the British Broadcasting Corporation aired a one-hour documentary on emerging viruses—in which Morse figured prominently.

At the same time, Morse was trying to create a stir among his professional colleagues. He co-authored a synopsis of the conference proceedings for the *Journal of Infectious Diseases* and edited the conference papers into a huge professional anthology, *Emerging Viruses.*

"I never thought a single conference would lead to all of this, whatever 'all of this' is," Morse says. But he seems genuinely pleased and flattered by the press calls he still receives, and somewhat overwhelmed by the stature he has achieved as the nation's most accessible spokesman on emerging viruses. Because of all the subsequent attention, he points out happily, he now calls the former president of his university "Josh."


The attention generated by Morse's conference came about in part because of his unconventional approach to setting up the program. Unlike most conference planners, he wanted to begin with a historical overview—not from another scientist, but from a historian. "It might have been because I was in love with a historian," he admits. At the time he was planning the conference, early 1989, he was romantically involved with Marilyn Gewirtz, a history teacher at the Bronx High School of Science, his own alma mater. Morse and Gewirtz had known each other casually years before and met again in the mid-eighties, after both of them had already been married and divorced. (They mar-

ried each other in 1991.) "But it's not only because I was in love with
a historian, but because I was in love with history, that I wanted to
be sure to have a historical perspective on this problem," Morse says.
So he called William McNeill, at the time an emeritus professor at the
University of Chicago, and asked him to be a speaker. McNeill agreed.
Next, Morse invited speakers from different branches of biology.
He looked for people who were thinking about pathogenesis, evolution, epidemiology, immunology, vaccines, genetics—all different
ways to try to tear apart the inscrutable virus. He invited Howard
Temin, American Cancer Society research professor at the University
of Wisconsin (under whom Morse studied while in graduate school at
Wisconsin), who won a Nobel Prize for his role in discovering the
enzyme that retroviruses use to reproduce; Bernard Fields, chairman
of microbiology at Harvard and the man who literally wrote the book
on viruses (the huge red tome *Fields' Virology* is in the personal library
of every medical student who can afford it); and Frank Fenner, the
Australian physician who led the myxomatosis campaign and is now
the world's leading expert on poxviruses. Morse included epidemiologists from the CDC; field virologists from the U.S. Army; entomologists, pathologists, veterinarians, zoologists. He put them all together
at the Hotel Washington and let them talk for three days straight.

His goal was to make the meeting not just interdisciplinary, but
intergenerational. He had envisioned this approach after attending a
session at the annual meeting of FASEB, the Federation of American
Societies of Experimental Biology, held in Las Vegas the previous
spring. On the first day, a small symposium applied a novel approach
to "Contemporary Topics in Immunology." For each topic under discussion, the organizer, Sheldon Cohen of the National Institutes of
Health, had paired an older scientist with a younger one. So for the
session on immunity to viruses, Jonas Salk, the developer of the polio
vaccine, head of the Salk Institute in San Diego, and at the time not
quite seventy-four years old, was invited to provide a historical perspective; and Barry Bloom, a fifty-one-year-old molecular biologist at
the Albert Einstein College of Medicine in New York, described his
work on developing vaccines using recombinant DNA technology.

"The pairing off was an inspiration," Morse recalls. "The old-
timers had been saying certain things for years, and no one was
listening; the younger ones, who concentrated more on genetic mechanisms, had not been thinking at all about pathogenesis, how infectious

agents cause disease." He arranged much the same thing for his own
conference, though not in such discrete pairs, combining the perspective of eminent epidemiologists who had been involved in tracking down
Lassa fever or Hantavirus with the views of younger molecular biologists
who understood these new virus diseases at the level of the gene.

The big surprise to come out of Morse's meeting was the discovery
that viral mutation was *not* the most important cause of viral emergence. For a mutation to present a real disease threat, it would have
to meet several criteria. It would have to provide some survival advantage to the virus, or at least be environmentally neutral, so the mutants would eventually outnumber the non-mutants. It would have to
be more lethal than its predecessor—otherwise the mutant would pose
no new hazard—but not so lethal that it killed its host before it could
get passed on to the next host. For a random mutation to meet all these
criteria is so unlikely as to be essentially unimaginable.

What does happen, though, is that human activities establish conditions in which the viruses that already exist are transported across
geographic or species boundaries. And this is what really captivated
Morse—and what led him, after the conference, to come up with a
whole new metaphor: viral traffic. "I like the viral traffic metaphor for
several reasons," he says. "To me it's a way of concentrating our attention on things that can be done." Morse is trying to move away
from the notion that new diseases arise de novo and therefore randomly
and unpredictably. He wants us to see that most new diseases are the
results of our own actions, and our actions can be changed. "Just as
with other kinds of traffic, viral traffic has its traffic indicators, Stop
and Go signals, and rules of the road."

Many of the environmental changes wrought by man's intrusion
into nature seem to be Go signals for emerging viruses. As we invade
the Amazon rain forest and urbanize the African savannah, we are
inadvertently putting people directly in the paths of viruses that otherwise would never have met up with human beings. In the last decade
or two, a variety of social, economic, and political changes in the ever-
shrinking global village has coincidentally created new roads along
which viruses can travel. These are not roads in the traditional sense,
but each serves as a conduit through which viruses and people can
move along and intersect, like marbles jettisoned along the wooden
pathways of an old-fashioned pinball machine. Among the man-made
mechanisms of viral conveyance are the following.

- Globe trotting, by people unaware they are harboring weird viruses, by animals imported from exotic locales for research purposes, or by virus-bearing mosquitoes or rats that stow away on transoceanic cargo ships.
- Urbanization, bringing huge populations of people and their virus-harboring animals to cities, to live under squalid conditions ripe for disease transmission.
- Global warming, providing a wider breeding ground for viruses previously confined to the tropics.
- Farming practices, which promote the recombination of viruses by placing in proximity animals harboring viruses that might be capable of infecting humans (like pigs and ducks) and by allowing them to exchange genetic material.
- Water management, which creates standing pools of water (like river dams or even irrigated rice paddies) that in turn encourage the proliferation of mosquitoes or other virus-carrying bugs.
- Modern medicine, which allows for new routes of virus transmission (intravenous tubes, organ transplants, blood transfusions) or reactivation (in patients whose immune systems are deliberately suppressed).
- Modern science, which can expose laboratory workers—and, subsequently, entire communities—to bizarre viruses under entirely unnatural conditions.

Urbanization, especially into the mega-cities of the developing world, has several effects on viral disease. When people from the countryside move to the city, they bring along some of the rodents that carry pathogenic viruses. Then the people settle, often into squalid surroundings, and the rodents proliferate. Mega-cities—cities of more than five million people—are notorious for their overcrowding, lack of plumbing, poor sewage systems, open trash heaps. All of these conditions allow for the spread of rodents, mosquitoes, and other disease vectors, as well as the pathogens they often transmit. And when city residents are poor, as so many are in the developing world, the accompanying stress and malnutrition can leave them especially susceptible to infection.

Hantavirus, the virus that first got Lederberg asking questions, emerged in the 1970s partly as a result of the rapid urbanization of Korea. The virus had long existed in the Korean countryside, mysteriously killing thousands of farmers and shepherds every fall. But its presence was just part of the background of disease accepted in many rural areas of Asia. Then Americans started getting sick. That's usually all it takes to get the attention of the world's best biomedical investigators. Ask a field virologist who has spent a lifetime on the African plains what constitutes an epidemic worth looking into, and he'll answer with characteristic cynicism, "The death of one white person." That's what happened with Hantavirus.

This was in the 1950s, when the United States and other members of a United Nations alliance were fighting in Korea. The U.S. Army noted that thousands of its troops were coming down with a flulike illness they had never had before. There were the typical high fevers and dehydration of influenza, but this illness also led, in severe cases, to internal bleeding, shock, and kidney dysfunction. Of some two thousand troops infected, several hundred died. The troops were suffering from what came to be called Korean hemorrhagic fever. (The name has since been changed to hemorrhagic fever with renal syndrome, which is both more medically precise and more politically correct.) Closely related diseases, all of them traceable to the same family of viruses and all of them involving the kidneys to some extent, occur not only in Korea but also in China, Russia and the other former Soviet republics, Greece, and Scandinavia. Together they account for about one hundred thousand cases, and five thousand deaths, every year.

At first, no one knew what caused hemorrhagic fever. "Many of the finest medical minds of the era were involved in the investigations," says James LeDuc, an epidemiologist who is now with the World Health Organization in Geneva and who was part of the research team. "But in spite of the massive effort invested and the tremendous knowledge gained in treatment and prevention of the disease, it was twenty years before the causative agent was finally isolated."

This happened in 1976, when a biologist at Korea University in Seoul, Dr. Ho Wang Lee, detected a new virus in the lung tissues of *Apodemus agrarius*—the so-called striped field mouse that exists throughout the countryside of Asia. He called it the Hantaan virus in honor of a nearby river; it is one of four related viruses that make up the genus known as Hantavirus. Lee's discovery explained how the troops were getting infected. War brought them out to the rice paddies where the striped field mouse lived. The mouse harbored the offending virus without getting sick, and shed it in its urine and feces. There the Hantavirus could remain for hours, perhaps even days, before finding a mouse or human host.

Soldiers in close contact with the ground were prime for infection. So were fieldworkers involved in the rice harvest. This helped explain a curiosity of the pattern of disease that had long been observed in Korea: hemorrhagic fever was most likely to strike in the late fall and early winter, and was far more likely to infect males than females. Once Lee isolated the virus and traced it to the striped field mouse, this mystery could be explained by Korean farming practices. Every fall the rice was harvested, and almost all of the workers involved in the harvest were male.

Within a few years, Lee discovered another surprise. Many residents of inner-city Seoul were coming down with a disease that looked like typical Hantavirus infection: fever, kidney dysfunction, internal bleeding, shock, and, in severe cases, total kidney failure. When he took samples of his patients' blood, he saw that they had antibodies to Hantavirus—an indication that they had been exposed to the virus and were mounting a response. But these patients were not farmers or soldiers; in many instances, they had never been outside the city limits. Somehow, they were being exposed to a rural virus without being exposed to its rural vector.

Like any good medical detective, Lee went to the homes of his patients to try to figure out how they got infected. They all lived in big apartment buildings, and it was easy for Lee to find, and catch, small rodents in the alleys nearby. He was looking for *Apodemus*, the striped field mouse, but all he found were common rats—the exact rat found in inner cities around the world. "They're known as wharf rats," LeDuc says. "They hang around the docks, hop onto ships, and make their way to port cities on every continent." The city rats, like their country cousins *Apodemus*, were also carriers of Hantavirus—a species that Lee named Seoul virus. Back in the lab, Lee found antibodies to Seoul virus in the rats' blood, and evidence of the virus itself in their organs. It seemed as if the creation of megacities had allowed the virus to leap from mouse to rat, and the poor living conditions in Seoul allowed the rat unusual proximity to human beings.

A secondary effect of urbanization is suburbanization, especially in the developed world. This inevitably leads to the invasion of wilderness areas to clear for new homes—which also plays a role in the creation of new viral diseases. As homes are built in regions that had been habitats for wild birds and animals, residents are exposed to animal viruses from which they previously were geographically removed. This proximity often leads to new human disease. The most widely publicized such disease is not viral in origin, but for a long time people thought it was. Lyme disease, a mysterious arthritis-like syndrome that first came to attention in the early 1970s, was clearly caused by *something* you could catch. Its pattern of transmission was the same as the one seen in many arboviruses (a shorthand for *ar*thropod-*b*orne viruses, the ones passed on by bites from a mosquito, tick, or flea). It clustered in certain regions, and its incidence increased in the summer months. At first it had seemed as though the source, whatever it was, was unique to the town of Old Lyme, Connecticut, where two young mothers had first noticed a high number of cases of arthritis among children in their bucolic neighborhood. But within a few years, it was clear that people throughout the northeastern United States were affected—most of them residents of towns like Old Lyme, where housing developments had sprung up in previously forested areas. The new housing brought deer into the backyards of suburbanites, placing people in proximity to the ticks the deer carried. At first, many scientists thought those ticks carried a virus. But when Lyme disease seemed to respond to antibiotics, that possibility was eliminated. Ultimately, it turned out that what was causing Lyme disease was a spirochete, another type of infectious microorganism, that is now known as *Borrelia burgdorferi*.

Viruses emerge, much the way this spirochete did, when animal habitats become suburbanized. Many of the vector-borne encephalitis viruses transmitted by mosquitoes that breed in tree holes, for instance, become a real threat only when humans get too close to forested regions, where they can get bitten by infected insects. Diseases such as eastern equine encephalitis and California encephalitis, quite rare in the United States, generally are restricted to locales where developers have stretched the boundaries of the suburbs farther and farther into the woods.

Compared to the suburbanized woodlands of the northeast, where some of the encephalitis viruses are found, the lush forests of the tropics are ablaze with viruses. And as the earth's average temperature creeps higher and higher, more of the world is turning into a tropic zone—meaning more of the world can provide a hospitable microenvironment from which new viruses can erupt. "Many tiny organisms that cannot endure freezing temperatures or low humidity thrive in tropical rain forests," writes William McNeill. "In the warmth and moisture of those environments, single-celled parasites can often survive for long periods of time outside the body of any host. Some potential parasites can exist as free-living organisms indefinitely." Inside the cells of many of these "potential parasites" are, no doubt, some viruses mankind has never seen before.

Living in this weird suspended animation, hovering in a soupy miasma of life-forms, is probably how tropical viruses have existed for centuries. Not really alive but not really anything else, they wait in the forest for a host to bump up against them and help them propagate. Many of them adapt a life cycle that is a perfect synchrony of needs: they spend some time in an arthropod, some time in a wild animal, and while the virus reproduces the arthropod and animal manage to stay healthy. For a long period of time, this equilibrium is sustained.

But as man builds roads into the forest and starts disrupting its delicate balance, the denizens of this fertile broth undergo a dramatic change. In the late 1950s, for instance, the Belém-Brasília Highway brought large numbers of human beings into the Amazon rain forest for the first time in history. "This started the cutting down of the rain forest, which the world is concerned about now," says Robert Shope, a Yale virologist who headed a Rockefeller Foundation laboratory in Belém, a northern Brazilian city of about half a million people at the time. "And when we took specimens from highway workers who were going into the forest and coming out again, we were able to isolate many viruses, some of them new to science." The mosquitoes and wild animals that Shope's colleagues captured along the new road were also carrying viruses no one had ever seen before. Until the highway was built, these viruses had undoubtedly been kept far away from people they might possibly infect—and from curious virologists who might try to identify them.

The next step for Shope and his co-workers was to determine if any of these new viruses were harmful to people. To do this, they needed to detect an outbreak of an actual disease. And that wasn't long in coming. One day in 1960, a Brazilian highway worker came to the municipal hospital in Belém. He had been healthy until two weeks before, when he developed a high fever and complained of headaches and confusion. Disturbing as his illness was, it seemed like an ordinary case of malaria, except for one symptom: he had lost all his hair. This made Shope examine the man's blood for weird viruses. Sure enough, he found one, a new virus named Guaroa virus in honor of the small town in Colombia where it had first been identified four years before. The Brazilian recovered. No other cases of hair-loss-and-fever were reported. The Guaroa virus now sits in a freeze-dried state in Shope's laboratory at Yale University, along with about five thousand vials of other exotic viruses awaiting a perplexing disease outbreak looking for an explanation.

As the earth's greenhouse gases raise the average international temperature—by as much as nine degrees Fahrenheit (or five degrees Celsius) in the next few decades, according to some predictions—the breeding ground for viruses will encompass an increasing portion of the surface of the earth. "Tropical diseases are going to move northward," says George Craig, an entomologist at the University of Notre Dame. "There's no doubt about it."

With global warming, we can also expect a proliferation of some of the insects that harbor and transmit viruses. Even a minor change toward warmer days, and longer summers, could spell trouble in terms of disease transmission. These changes are likely to lead to a disproportionately large increase in the population of the most common vectors of deadly viruses, such as mosquitoes.

"A very small amount of global warming can change the synchrony between development processes of an insect and its predator," says Richard Levins, an ecologist and mathematician at the Harvard School of Public Health. "Insects that now mature at the same time will mature at different times" if the average annual temperature of a region increases by even a single degree, he says. "Predators may lose synchrony with their prey."

Levins develops mathematical models for predicting the disease impact of environmental changes such as global warming. Here is how it would work, for instance, with the relative populations of the *Culex pipiens* mosquito—which transmits St. Louis encephalitis—and its natural predator, the dragonfly. The central notion is the one of degree-day—a day that achieves an average temperature that is one degree higher than a predefined threshold. (If the average temperature is two degrees higher than the threshold, that day counts as two degree-days; three degrees higher counts as three degree-days.) In order to reach maturity, a larval insect has to accumulate a fixed number of degree-days. The number needed before hatching, and the temperature threshold that determines when a day counts as a degree-day, differ from one species to another but are constant within the same species.

Suppose that for *Culex* the threshold for a degree-day is fourteen degrees Celsius. Say the mosquito must accumulate fifty degree-days to develop. (These are not the actual numbers; Levins pulled them out of thin air to illustrate his point.) If the average daily temperature in April and May is fifteen degrees, the mosquito accumulates one degree-day every calendar day, and after fifty calendar days it reaches maturity. Now let's say global warming increases the average April temperatures to sixteen degrees. This doubles the rate at which degree-days are accumulated—two degree-days each day instead of one. So the mosquito reaches maturity after just twenty-five calendar days have passed.

Now imagine that *Culex*'s natural predator, the dragonfly, also comes to maturity in fifty days. But this schedule is determined by different factors. The dragonfly has a degree-day threshold of *thirteen* degrees, and must accumulate *one hundred* degree-days to reach maturity. When the actual temperature for April and May is fifteen degrees, this works out fine: with two degree-days accumulated every day, it hatches in the same fifty days it takes its prey to hatch. But when global warming increases the temperature to sixteen degrees, the predator and prey get out of synch. "When you raise the temperature to sixteen," Levins says, "that means the predator has increased by fifty percent in its rate of accumulation of degree-days." But the mosquito has doubled *its* rate. "Whereas both of them took fifty days to develop before, now one of them is down to twenty-five days and the other to about thirty-five days. The synchrony is broken." The result: young mosquitoes are let loose without a predator, and the number of humans bitten, and possibly infected with St. Louis encephalitis, can increase significantly.

The pollution of our atmosphere leads to more environmental catastrophes than the greenhouse effect. And many of these other problems will also have a public health impact. The destruction of the earth's protective ozone layer, for instance, could lead to a disruption in immunity in every creature on earth. The ozone layer keeps the harmful ultraviolet rays of the sun from reaching the earth's surface, and without it we will be exposed to far more ultraviolet radiation. When that happens, immunity can be deranged. Scientists have found, for instance, that experimental mice exposed to ultraviolet radiation have lower than normal numbers of immune system cells in their bloodstreams.

"It is a whole series of interconnected problems," writes Thomas Lovejoy, assistant secretary for extramural affairs at the Smithsonian Institution. Agriculture in the developing world will also be hurt by climate changes, he predicts, compounding the nutritional hardship that already exists. The stresses of poverty, complicated by pollution-induced problems with nutrition and immunity, will lead to what Lovejoy calls "a nastier epidemic situation."

Other viral diseases have gained a foothold because of the agricultural whims of "modern" man. The deadly Argentinian hemorrhagic fever, for instance, emerged after World War II, when the pampas grasses of Argentina were cleared to make way for new farms in the north-western provinces of the country. When the tall grasses disappeared, so did predators of the species of rodent that carried the Junin virus. Beginning late every summer (which is February in Argentina), and ending early in winter, the rodent populations of *Calomys laucha* and *Calomys musculinus* increase. The human population increases, too, in these months, as transient farmers move in for the harvest. And every year there is an outbreak of a terrible infection that kills up to 20 percent of its victims.

The symptoms are devastating: fevers up to 104°F (40°C), chills, headache, eye and muscle pain, vomiting, low blood pressure, dehydration, and, in the worst cases, signs of internal hemorrhaging, such as bleeding gums, nosebleeds, and blood in the urine and vomit. When death comes, it is usually from the profound shock caused by leakage of plasma—the liquid portion of the blood—out of the blood vessels and into surrounding tissue. Those at highest risk are male, between the ages of twenty and forty-nine, especially those employed as agricultural workers.

So that's one way farming practices can affect the kind of viruses that are passed on to people: by determining where and when farm-workers are exposed to virus-carrying animals. This explains the seasonal outbreaks, usually among people harvesting crops, of such disparate diseases as Bolivian hemorrhagic fever in Bolivia, hemorrhagic fever with renal syndrome in Korea, and Rift Valley fever in Egypt. But there's another way farming leads to viral emergence. Certain farming practices actually put two or more *animal* species in proximity, and they exchange the viruses that each carries. When the viruses mix it up in one or the other animal, they may exchange bits of genetic information and end up as totally changed viruses—viruses that are capable of infecting a third animal, often humans.

That is how influenza probably is given an opportunity to change. "All the genes of all influenza viruses in the world are being maintained in aquatic birds," says Robert Webster of the St. Jude's Children's Research Hospital in Memphis. Wild ducks and other waterfowl carry influenza viruses in their bellies; the ducks are not at all affected. The intestinal tract is an ideal place for the virus, allowing it to replicate with ease and giving it a handy way to migrate to the next host. The new viruses pass out in the bird's feces and contaminate the water where other ducks swim. When uninfected ducks drink the water, the virus passes right into a new intestinal tract.

All goes well as long as the duck community stays relatively intact. Once the ducks start mingling with other animals, though, the influenza viruses they carry can mingle with whatever viruses the *other* animals carry. Here is where farming practices come in. Another "reservoir" for influenza viruses happens to be the domestic pig, and many farms, especially in Asia, mix pigs and ducks deliberately. Now it's the pigs that drink virus-infected waters, and inside the pig's body two strains of influenza virus are placed side by side. They often exchange bits of genetic material, emerging as a new flu strain that can be devastating to a human population that has no immunity to it.

Just as pigs and ducks can serve as viral mixing vessels in the fields of China, so can certain human beings become mixing vessels in the hospitals of America or any other industrialized nation. Modern medicine has created a population of deliberately immunosuppressed patients, who take drugs to render their immune systems helpless in the face of pathogens. These are patients for whom a functional immune system, for one reason or another, would run counter to their body's best interest.

Most of the people who receive chronic doses of cyclosporine or other immune suppressors are those who have received organ transplants. With their immune systems deliberately impaired, their natural mechanisms for rejecting foreign tissue are subverted. This is good for the success of the transplanted organ—it keeps their bodies from rejecting it—but it can lead to some dreadful side effects, such as kidney and liver toxicity. Indeed, the very act of taking immune suppressors can sometimes lead to chronic illness of its own, in particular hypertension and cancers. But for transplant patients who would die without the foreign organ, and who would never tolerate the new organ without the drugs, the trade-off is worth it.

A smaller group of people takes these powerful drugs to treat autoimmune diseases, such as multiple sclerosis, lupus, and diabetes. In autoimmune disease, the body misidentifies its own tissue as foreign and tries to reject it, much the way it would reject a transplanted kidney. When the immune system tries to fight off particular nerve cells, multiple sclerosis results; when it tries to fight off certain cells in the pancreas, the result is diabetes. In some medical centers, people with autoimmune diseases are being treated with the same powerful immunosuppressant drugs that have been used for years in transplant patients. In addition, many of the chemotherapeutic agents used to fight cancer have the unintended side effect of depressing the patient's immune system.

The result of all these deliberately, or inadvertently, immunosuppressed patients could be a new class of human mixing vessels for viruses. These people become infected so easily that it's quite likely they will, on occasion, catch two or more infections simultaneously. When that happens, if the viruses are especially good at recombining—the way the influenza viruses are—a whole new strain could emerge from the damaged person's body and start infecting an entire community. In the face of a recombined new virus, we are all nearly as helpless as immunosuppressed people are. "I can't think of any situations in which this has actually happened," says Morse. "But it's certainly theoretically possible."

Also possible is the prospect of a previously harmful virus's mutating in immunosuppressed patients in a way that renders it suddenly pathogenic. In one scenario, a middle-aged woman receives a kidney transplant, and she begins taking cyclosporine. In her new kidney was a latent cytomegalovirus (CMV), a common herpesvirus that usually lies dormant in people after causing an initial active infection. The kidney donor was not aware of harboring CMV; in all probability, the infection never would have become reactivated in the donor's lifetime. And in the kidney recipient, the CMV ordinarily would be just as innocuous; indeed, she may well be carrying a latent CMV infection of her own. But because of the cyclosporine she is taking, the recipient's body cannot keep the new CMV in its latent state; for a virus to remain latent, the immune system itself must impose a certain restrictive control, usually by continuous surveillance by the marauding cells called macrophages. The CMV starts replicating again, in the patient's new kidney. "I'm concerned that the chances for mutation increase with the number of generations," says Edwin D. Kilbourne, a virologist at the Mt. Sinai School of Medicine in New York. The patient's immunosuppression, he says, "provides increased replication possibilities. If the virus has a chance more and more to flourish in the tissues of the adult, then it's possible that new tropisms can emerge, and new virulence, and new transmission patterns." In short, CMV can turn into a virus that is quite dangerous for adults, rather than a herpesvirus that has reached a state of equilibrium with its human host.

Other miracles of modern medicine have coincidentally created new ways to introduce viruses into the body. Doctors can now treat deficiency diseases with human hormones, replace lost blood during surgery with blood transfusions, repair damaged organs with transplants taken from donors. In each case, the hormone or blood or organ might be harboring a virus that is so weird and so hidden that there's no good way to test for it.

Among the biological products that have turned into a viral time bomb is human growth hormone. Derived from the pituitary glands of human cadavers, this product was inadvertently contaminated with the virus that causes Creutzfeldt-Jakob disease, a strange slow virus that can persist unnoticed for twenty years or more. It is thought to be related to the viruses that cause mad cow disease, now raging through England, and kuru, the neurological disease common in New Guinea that D. Carleton Gajdusek studied in the 1950s. An associate of Gajdusek's, Clarence J. ("Joe") Gibbs of the National Institute of Neurological Disorders and Stroke, has studied unusual paths of transmission of Creutzfeldt-Jakob disease, a relentless and fatal brain degeneration. Human growth hormone is among the most unusual. At least fifteen deaths from Creutzfeldt-Jakob have been caused by human growth hormone. But this may be only the beginning, he says, "since our estimates are that thirty thousand young people would have received this treatment worldwide." Gibbs and his co-workers recommended that cadaver-derived growth hormone be taken off the market, and it has since been replaced by a synthetically made hormone that carries no risk of infection.

Other iatrogenic (physician-caused) sources of Creutzfeldt-Jakob virus also have been documented. In at least one case, a woman received a cornea from a man who had died of other causes, never knowing he was harboring the slow virus in his optic nerve tissue. The optic nerve's contamination apparently had spread to the man's eye tissue as well. The woman who received his cornea died, eighteen months after her sight was restored, of the horrible neurological disease. The same virus has also been transmitted through even more sophisticated medical techniques. In two cases in Switzerland, two young people being treated for epilepsy by the implantation of silver electrodes placed deep in their brains were also inadvertently given a direct dose of Creutzfeldt-Jakob virus; the electrodes had presumably been used first on a patient harboring the virus, either symptomatically or asymptomatically. The Swiss cases involved a seventeen-year-old boy and a twenty-three-year-old woman; both died less than two years after their electrode therapy. In at least five other cases, surgical instruments that had been cleaned—as the brain electrodes were—with heat disinfection, alcohol, and formaldehyde nonetheless still carried the virus, and the patients on whom the instruments were used all died of Creutzfeldt-Jakob disease within about two years of surgery. And in five more cases, an implant of cells from the dura mater—the outer membrane covering the spinal cord and brain—also led to transmission of Creutzfeldt-Jakob disease and a quick, nasty death for the recipients.

The sudden appearance of symptoms of a viral infection in a small cluster of people helped Gibbs and others trace the source of the Creutzfeldt-Jakob virus in all these unfortunate individuals—even though the disease, in these cases, was far more virulent and speedy than the ordinarily languorous degeneration. Theoretically, receiving a direct dose of virus through an organ transplant or a contaminated instrument makes for a meaner clinical course. In a similar way, the outbreak of a bizarre viral syndrome in West Germany in 1967 received immediate attention because the cluster of people was so sharply defined, and their illness was so dramatic.

It began in a town called Marburg. Twenty-five people suddenly developed a mysterious set of symptoms: high fever, bloodshot eyes, a rash covering their bodies, bloody vomiting, and diarrhea. In seven of them, the bleeding became uncontrollable; they died. It wasn't really hard to make a connection between the patients who were taking sick. All the victims were employed at the same place: a research laboratory that was manufacturing polio vaccine. The trick was finding what the sick workers had in common, and what made them different from people who worked elsewhere in the facility.

"It was pretty obvious," says Joe McCormick, the epidemiologist from the Centers for Disease Control who was called in to explain the outbreak (much as he would be called, some twenty years later, to investigate the case of Lassa fever in Chicago). "It was a large outbreak; you were out of the endemic area [the region where the virus occurs in nature], so there was only one possibility." All of the victims were working with cultures of monkey cells, specifically kidney cells derived from vervet (green) monkeys caught the previous month in Uganda. Monkey kidney cells were the only animal cells in which poliovirus could grow.

By the time the outbreak was over, one month after it began, a total of thirty people had been infected—not only in Marburg but also in Frankfurt and in Belgrade, Yugoslavia. The common denominator was that in each locale, laboratory workers had been using tissues from green monkeys imported in the same shipment from the same African dealer. But the pattern of infection was a curious one. Animal handlers—who cleaned the monkeys' cages, fed them, groomed them—did not get infected. The only workers who got sick were those who surgically removed the monkeys' organs or who worked with those organ cells in culture. And in five cases, the close contacts of these workers—wives, girlfriends, hospital attendants—were infected too. The epidemic was quickly contained by isolating all patients and killing all suspect monkeys. Seven of the thirty infected people died; the rest recovered with no residual effects. And then, as quickly as it had erupted, the disease just disappeared.

Since the original outbreak, the virus—now called Marburg virus—has infected human beings only twice. The first occurrence was in a young Australian hitchhiking in 1976 through what was then Rhodesia; the second was in a Swedish tourist who got sick during a vacation in Kenya in early 1990. The Australian hitchhiker died, and his girlfriend and the nurse who treated him caught the disease and recovered. The Swedish tourist spent two weeks in the intensive care unit of a Stockholm hospital before going home. Today, the Marburg virus sits in a freeze-dried state in just a few laboratories around the world. The 1967 outbreak proved the virus so highly contagious, and its effects so deadly, that in all this time no lab has been allowed to investigate it more thoroughly to find out how it works—even when it is frozen.

It seemed a unique and purely exotic story, to relegate to the scientific curiosity shelf. But for Stephen Morse, who was still in high school when the virus was discovered, the Marburg affair is a cautionary tale. Indeed, it's just the sort of dreadful cascade of events that Joshua Lederberg envisioned when he passed on the query about Hantavirus: a virus resides without harm in a wild animal, the animal is caught and its tissue propagated for use in the lab, and the resident virus is released to the environment. And this cascade occurred again, with a few new wrinkles, as recently as 1989, when two shipments of monkeys from the Philippines came into a Virginia holding facility and were found to be harboring a variant of an especially deadly virus known as Ebola.

Fortunately, lab outbreaks generally stop after the first or second generation of contacts. The viruses are so strange, and so fragile, that they have so far proved ineffective at being transmitted person to person. But a virus does not have to be especially efficient in order to cause a mini-epidemic right inside the laboratory or hospital. Scientists work with viruses in high concentrations, and they manipulate them in ways that make transmission especially likely, performing autopsies of diseased animals, sucking up virus-laden solutions through glass pipettes. William McNeill has said that in a new plague, "the doctors would be the first to go"; a corollary to this is that very early in the plague's progression, the others dying off would be the scientists investigating its cause.

In theory, once the virus takes up residence in its new host—the physician or the virology lab worker—it can begin to adapt to humans. This is an important point, and one worth returning to: the rapid mutation rate of most viruses, while no longer thought to be responsible for the sudden creation of new strains, *is* responsible for the virus's ability to adapt itself speedily to a new evolutionary niche. Once the new niche presents itself—in this case, in the form of bringing animals or animal cells that harbor viruses into direct contact with lab workers—the virus may subsequently be quite capable of grabbing on to a new opportunity and adapting itself to a new host. Viruses and other microbes "thrive on undercurrents of opportunity," says Richard Krause of the National Institutes of Health. "They have also been great innovators who have developed new powers of pathogenic vigor through genetic versatility."

The result of this rapid pace of evolutionary adaptation could be that the virus that began as a lab outbreak could soon become more and more efficient at infecting human beings. Should that happen, once scientists bring this new infection into the community, no one would be immune.

The story of virology is not simply one of new and menacing threats lurking around every molecular corner. Just as human manipulation can be blamed for the creation of new threats to our health, so can it be praised for actually eliminating certain viral disease. And just as viruses emerge over the course of evolution, so do they—albeit rather rarely—disappear.

The most obvious instance of human intervention wiping out a virus was in the case of smallpox, once considered the most communicable of all infections. This horrid disease, which caused weeping sores all over the body and killed more than half its victims, was completely eradicated after a worldwide vaccination program. The last case of "wild" smallpox was recorded in October 1977; the last non-military smallpox vaccination was given in 1982. For a long time, the only possibility of smallpox transmission was in the laboratory. The last two outbreaks of smallpox in Europe both occurred in England, and both resulted from laboratory exposure. Three people were infected in London in March 1973 after a medical technician contracted smallpox from the virus he was handling and passed it on to two others; the two secondary cases died. In August 1978, an even weirder outbreak occurred, involving a female photographer taking pictures in a medical laboratory in Birmingham, England. When she contracted smallpox, medical detectives traced it back to the lab in which she had worked; it was directly connected, via a ventilation duct, to the smallpox laboratory of prominent scientist Sir Henry Bedson. The photographer died, and her mother also developed a case of smallpox, from which she recovered. Sir Henry, though, did not recover; overcome with remorse over the young woman's fatal case of smallpox, Bedson committed suicide.

After these tragedies, the governments of the NATO allies and of the then–Soviet bloc agreed to limit smallpox stockpiles to just two locations: the high-containment laboratory at the Centers for Disease Control in Atlanta and the Research Institute for Viral Preparations in Moscow. They are now mapping every gene on the smallpox virus, in the hopes of storing the sequence on a computer disk and ultimately destroying even these last remaining vials.

Sometimes, it seems, viruses disappear by themselves. In the fifteenth and sixteenth centuries, Europeans were wracked by a mysterious disease known as sweating sickness, or the English Sweate. Viruses weren't even discovered yet—they wouldn't be imagined until 1898, nor visualized until 1939—but the symptoms of the disease sound a lot like conditions we know today to be caused by viruses. "Victims died quickly, usually only twenty-four hours after the onset of a brutal fever and a drenching, stinking sweat," notes Mirko D. Grmek, a Yugoslav physician and medical historian who now teaches at the Sorbonne. The Sweate appeared suddenly in 1485, erupted five times in England and France, and disappeared in 1551. Contemporary experts believe that whatever caused the Sweate was probably a virus, and that whatever virus it was is now extinct. If that is so, it means that some of the viruses we're plagued by today may eventually run themselves into extinction too—either because they kill their hosts too quickly, or because they lose their ability to pass from one host to the next.

The paths of viral evolution have occupied the spare hours of many of today's experts on emerging viruses. Richard Krause often invoked viral evolution in essays and speeches over the years. "The reemergence of old problems in new garments is programmed in the genetic machinery of evolution," he wrote in 1978. "In mapping a strategy for the control of infectious diseases, therefore, we must be . . . alert to the sovereignty of the evolutionary tide." Stephen Morse, from the perspective of a scientist trained during the dawn of molecular biology, puts a genetic twist to his musings about evolution. "The development

THE TINIEST MENACE

of methods using molecular sequence comparison to construct phylogenetic, or family, trees is one of the most powerful and valuable applications of molecular evolutionary theory, and has been most useful for elucidating relationships among viruses," he has written. Since the advent of molecular biology, he adds, the study of viral evolution is now "almost entirely a field of molecular comparisons."

And then there is Edwin Kilbourne, the Mt. Sinai virologist who became a lightning rod for controversy because of the leading role he played in sponsoring the ill-fated "swine flu" vaccine in 1976. Several years later, Kilbourne—a tall, gaunt man with a white goatee who, in his white lab coat, is a cross between Pete Seeger and Jonas Salk—was asked to write a paper about the epidemiology of gene-altered viruses. Public sentiment at the time was focused on trying to restrict recombinant DNA research and dangerous mutants that might escape from the laboratory. Having sharpened his public relations skills in the long months of early 1976, defending his recommendation to go ahead with the vaccine for an epidemic that never materialized, Kilbourne could scarcely resist this new, fiery controversy. He accepted the journal editor's assignment and managed to mark it with his own wry brand of humor. He imagined a totally new type of virus, which he mischievously called the Maximally Malignant (Monster) Virus. (In a subtle jab at scientific conventions, he referred to it throughout the paper by its initials, MMMV, just the way an entirely serious paper would.) This dreadful virus, he wrote, would have the most lethal elements of all viruses currently known: "the environmental stability of poliovirus, the antigenic mutability of influenza virus, the unrestricted host range of rabies virus, and the latency or reactivation potential of a herpesvirus." It would be transmitted through the air, like influenza, and replicate in the lower respiratory tract, again like influenza; it would cause immune suppression in survivors and lead to frequent flare-ups of infection. And it would, like the AIDS virus, be a retrovirus that manages to insert its own genes directly into the host's cell nucleus.

The MMMV does not exist, of course, and it's difficult—though not impossible—to imagine a human mentality vicious enough to try to create one artificially for use as a biological weapon. Kilbourne invokes MMMV primarily to make a point: that with viruses, in which only a few changes can make a huge difference in the way the microbes behave, trying to predict the paths of evolution and emergence can be a treacherous affair indeed.

# 2

# Case Study: Why AIDS Emerged

In December 1977, a Danish surgeon named Margrethe Rask died in Copenhagen of three unusual infections: *Pneumocystis carinii* pneumonia, *Candida* in the mouth and throat, and septicemia (blood poisoning) with *Staphylococcus albus*. She was forty-seven years old, and had been sick for nearly two years. From 1972 through 1977, Rask was a field surgeon in Africa, working first in a rural hospital in nothern Zaire and then in a larger hospital in Kinshasa, the capital.

In October 1982, Claude Chardon, a French geologist, died in Paris with a *Cryptosporidium* infection of the intestinal tract and a *Toxoplasma* infection of the brain. He was thirty-five years old, married, the father of a young girl; he had been sick for more than a year. His death followed by four years a serious car crash he was involved in while doing fieldwork in Haiti. After the crash, in 1978, Chardon's left arm was amputated, and he was saved by emergency transfusions of fresh blood donated by eight generous Haitians.

In March 1984, a Canadian flight attendant died in Quebec of Kaposi's sarcoma (a rare and usually slow-growing cancer) complicated by *Pneumocystis* pneumonia. The man, Gaetan Dugas, was thirty-one years old, had had cancer since 1980, and traveled constantly—as a flight attendant, he could fly for free—throughout the United States and Canada, to France, to the Caribbean, encountering dozens of male sexual partners at every stop.

With the benefit of hindsight, we see clearly now that Rask, Chardon, and Dugas were all early victims of AIDS. They all died with the opportunistic infections that have become the hallmark of full-blown AIDS: infections by organisms that would cause no problems

37

to people with functional immune systems, but that proliferate in AIDS patients who lack all defenses against them. Among the most common opportunistic infections that have bedeviled AIDS patients are the ones that appeared in these early cases: *Pneumocystis carinii* pneumonia (so common now that it's known by its shorthand, PCP), toxoplasmosis, cryptosporidiosis, Kaposi's sarcoma. Other opportunistic infections also took hold during the first decade of the AIDS epidemic: *Mycobacterium avium-intracellulare*, cytomegalovirus retinitis, oral candidiasis, and perhaps a dozen more.

The details of the cases of Rask, Chardon, and Dugas—and the details of the cases of thousands of others—were what helped scientists finally to formulate a cohesive theory about where AIDS started, and how it could unleash such ferocity on people all across the globe. Rask might have helped bring the AIDS virus out of Africa and into Europe. Chardon might have helped add to the ever-more-complex web of transfer when he moved, with his blood-borne infection, from Haiti to France. And Dugas might have served as the North American vortex from which the furious storm of viral spread began. Some American epidemiologists called Dugas "Patient Zero," the public health equivalent of ground zero, which indicates the spot where an atomic bomb drops. He is known to have caused at least forty cases of AIDS in New York, Los Angeles, and eight other American cities, either by having sex with the patients themselves or by having sex with men with whom the patients also had sex. No one knows how many more men he infected around the world who could not be traced. Dugas was terrifically handsome, flirtatious, seductive—by his own count, he had about 250 sexual partners every year, even after his illness began. Dugas was also quite vindictive, using his infection like a deadly weapon. Even after being told, in late 1982, that unprotected sex with him would endanger his partners, he never warned people that he might give them AIDS. Instead, he seemed to look for conquests with a frenzied urgency, telling men only after the act was completed, "I'm going to die and so are you."

In many ways, AIDS serves as a perfect example of viral emergence in action, and close study of the epidemic's earliest rumblings helps uncover patterns and truths about why viruses emerge. Indeed, of the many human activities now known to contribute to the emergence of new viruses, at least half a dozen came into play in the case of AIDS: globe-trotting, migration to the cities, behavioral changes, techniques of modern medicine, gene recombination, and laboratory manipulations have all contributed to the emergence and the spread of AIDS.

AIDS changes the stakes in the study of emerging viruses. With the world now in the grip of a deadly virus that recently emerged, the theories about where new viruses come from—theories that tropical disease experts and epidemiologists have been batting about for years—suddenly take center stage. To the people in the field, this is at once invigorating and terrifying. These scientists have not been groomed for public relations. They usually prefer the hardships of life in a remote African village, or the difficulties of working with dangerous pathogens using test tubes as thin as a human hair, to the glare, noise, and controversy of being in the public eye.

Stephen Morse, for instance, knows that the existence of AIDS, an epidemic that has captivated the public like no medical condition since polio, is partly responsible for whatever interest has been generated by his meetings, articles, and talks. "Obviously, people are as interested as they are in this question at this time because they want to know where AIDS came from," he says. If not for AIDS, his 1989 conference on emerging viruses might never have received federal funding, and its conclusions would have been of only passing intellectual interest. If not for AIDS, the National Academy of Sciences would never have established a blue-ribbon panel on emerging microbes, on which Morse served as a subcommittee chairman and which allowed him to get to know as intellectual peers the laureates among biomedical researchers, men and women nearly twenty years his senior.

Another man might feel rather clever, or at least extraordinarily lucky, to be able to capitalize on such brilliant timing. But Morse seems somewhat saddened by it all, conscious of the gravity of the enterprise he is now engaged in. If he becomes a major spokesman on the subject of emerging viruses—as close to "famous" as a virologist is likely to get—he knows it will have been because of the suffering and death of hundreds of thousands of people with AIDS. Of course, an argument can be made that if he is successful in setting up a mechanism for predicting "the next AIDS" in time, he may eliminate quite a good deal of suffering from diseases we can only imagine. But he is, nonetheless, still benefiting—quite unintentionally—from the existence of AIDS.

Morse is a kind person who makes small talk with everyone he runs into, from the woman dishing out macaroni and cheese at the Rockefeller University cafeteria to the Nobel laureate sitting across the dining hall, from the security guard at the York Avenue gates to the department chairman intercepted in the mail room. It's unlikely that someone so attuned to other people could overlook the fact that in modern biomedical research, scientists' reputations are often made because huge masses of ordinary people are in pain.

"Rudely jolted by AIDS back into an awareness of infectious diseases, we now find ourselves in a period of great uncertainty, poised for the AIDS of the future," Morse wrote in an essay on the history of AIDS. "We cannot help but wonder what other catastrophes are waiting to pounce on us."

For Richard Krause, there is perhaps a different undercurrent of gloom when he speculates about the origins of AIDS. Krause was in a position of authority early in the AIDS epidemic, when a handful of epidemiologists in California and New York were begging the National Institutes of Health, including the institute that Krause headed, to take the lead in studying the perplexing outbreak. Krause subsequently came under fire for being part of NIH's collectively slow response after the deaths of the first few dozen gay men were documented. He left NIH in 1984 for reasons he says had nothing to do with AIDS. "When the disease struck in 1981, there were just a few cases," Krause says now in his own defense. "If we jumped through hoops every time there were a very few cases of this and that, we wouldn't be keeping our eye on the ball." Even if NIH had reacted more quickly, Krause says, the course of the epidemic would not have changed. "I don't see how the [scientific] work could have moved any faster. Within a year and a half we knew that AIDS was sexually transmitted; that's very tough [to establish] for a disease with that long a latent period, very tough. And then within three years the virus had been identified. Six months after that there was a blood test. Two years later there was a drug." Krause believes these developments to have been speedy enough.

In the middle of the grip of AIDS, scientists and non-scientists alike lack the cool objectivity needed for an accurate historical perspective. Indeed, because we are talking here about a sexually transmitted disease associated with behaviors that break the rules of straitlaced morality, we may always lack objectivity about AIDS, no matter how far removed we are in time. (No one ever figured out exactly why syphilis erupted in the late 1400s; the French called it "the disease of Naples," and the Italians called it "French disease." After centuries of finger-pointing, the issue of how syphilis originated simply evaporated, subsumed by other more pressing medical uncertainties.) Early on in the AIDS epidemic, in 1983 and 1984, religious fundamentalists were quick to call it the "wrath of God," divine punishment for people engaging in indiscriminate homosexual sex or rampant drug abuse. That attitude has, mercifully, diminished since then, though it still manifests itself occasionally. The "wrath of God" mentality, for instance, might be behind the U.S. government's continuing policy— against the advice of all public health experts and contrary to the practices of any other country in the world—to bar any visitor or immigrant who tests positive for the AIDS virus.

But it's not just religious zealots or political conservatives who turn spiritual when talking about AIDS. Even among scientists, analysis of this disease can lead to some distinctly non-scientific perspectives. An edge comes into some people's voices, a prurient eagerness to talk about the details of male prostitution or sex with monkeys, a willingness to blame another nationality or another ethnic group for a disease that is too dreadful to imagine originating closer to home. Often, these tendencies interfere with the ability to think clearly. How else to explain all the theories still in circulation about how the AIDS virus first made the leap—if indeed such a leap was even required—from monkey to man?

The most widely accepted theory about the origin of AIDS holds that the AIDS virus (human immunodeficiency virus, or HIV) began as a nonhuman primate virus (simian immunodeficiency virus, or SIV) that for some reason made a cross-species jump and began infecting people. The timing of that jump is still the subject of much debate. Analysis of stored samples of serum from patients throughout Europe, Africa, and North America early in this century seems to indicate that an SIV-like virus began infecting humans no less than fifty, but no more than one hundred, years ago. But the location for that jump from monkey to man is less controversial: most leading virologists locate the first stirrings of AIDS on the African continent, the region where mankind itself originated. Since there are two very different strains of HIV, it is possible that the cross-species leap occurred at different points in time. The strain more common in western Africa, known as

HIV-2, is so similar to SIV as to be almost indistinguishable, leading many American and European scientists to believe that it made its leap relatively recently, perhaps fifty or sixty years ago. The strain known as HIV-1, which is more common in the rest of the world, is now thought to have begun infecting humans at least two generations earlier, giving it time to evolve in its new host species and to diverge more and more from its animal origins.

In the case of both HIV-1 and HIV-2, conventional wisdom has it that something happened in the early 1970s to turn an already-simmering virus into a public health force to be reckoned with. This "something" could have involved changes in the virus itself (though this is unlikely), or it could have involved changes in the human behavior that facilitated its spread, both within Africa and from Africa to countries halfway around the world.

But according to another theory about the origin of AIDS, the pandemic was not seeded in Africa, and the first eruptions of trouble can be traced back much earlier than the 1970s, at least as far back as the 1930s and 1940s. According to this theory, the AIDS virus never changed at all, nor did human behavior. What really changed was the global burden of *other* infectious diseases. As other diseases were conquered, the effects of HIV could emerge from behind their camouflage. Mirko Grmek of the Sorbonne coined the term *pathocenosis* to describe this view that mankind is plagued by a constant burden of infectious diseases. Whenever one group of infections is eliminated, he writes, a new group rushes in to fill the vacuum. Intriguing as it is, though, this theory is resisted by many scientists, especially Americans, in part because it places some of the world's earliest signs of AIDS right in the United States. It also carries a touch of fatalism, a philosophy that runs counter to the American can-do approach to thorny problems.

Theories about the African origins of AIDS are based largely on serological studies of monkeys living in the westernmost countries of the continent: most such monkeys show signs of having been infected with SIV, the simian retrovirus that looks almost identical to the human retrovirus HIV-2. Up to 70 percent of monkeys in western Africa are infected with SIV; up to 15 percent of healthy adults in the same region test positive for HIV-2. And for high-risk populations, the infection rate skyrockets: among female prostitutes in Bissau, Guinea-Bissau, for example, the rate of HIV-2 infection is an incredible 64 percent. Discovered in 1985, HIV-2 is a close cousin to the AIDS virus (which is designated HIV-1 because its discovery came two years earlier). It is endemic to Guinea-Bissau, Senegal, Gambia, and the Cape Verde islands. But it hardly exists at all in much of central Africa, such as Zaire, Kenya, Burundi, Congo, Uganda, and other countries where AIDS is so prevalent. Nor is it found on the other continents where AIDS is most entrenched: North America, South America, and western Europe.

HIV-2 might be closely related to HIV-1, but what it really resembles, almost to the point of identity, is the monkey virus SIV. The genetic resemblance between HIV-2 and SIV is so close that their different names can be accounted for only by scientific convention. "The fact that we call monkey viruses SIVs and we call human viruses HIVs reflects nothing more than the fact that the human ones are in people and the monkey ones are in monkeys," says Max Essex, director of the AIDS Institute at the Harvard School of Public Health and co-discoverer of SIV. "If you went to a different species of monkey that hasn't been very well studied—say a mandrill monkey—and you found one of these viruses because the monkey had antibodies that cross-reacted with other SIVs, and then you isolated that virus and genetically cloned it and sequenced it, and then you sent it to the gene sequence experts under code and said, 'Is this a monkey virus or a human virus? I don't know,' they would send it back and say, 'We don't know, either.'" In other words, the order in which the viral gene constituents appear—which directs everything about the way the virus functions—is almost exactly the same in the monkey virus as in the human virus. Indeed, different strains of SIV taken from different sooty mangabeys (a type of monkey that lives in the coastal forests of West Africa) are more similar to HIV-2 than they are to each other. "And there's every reason to assume that HIV-2 can infect monkeys," says Essex, "and that at least some SIVs can infect people."

Such genetic similarity is a sign of a common heritage. When scientists try to draw inferences about the evolutionary relationships between any two species—frog and turtle, tiger and lion, monkey virus and human virus—they look at regions of their respective DNAs that are the same, or homologous. The greater the percentage of homology, the more closely related the two species are on the phylogenetic tree, the family tree of living things. Recently, a colleague of Essex's, Ronald Desrosiers of the New England Regional Primate Center in South-borough, Massachusetts, reported an 85 percent homology between HIV-2 and the type of SIV found in sooty mangabeys. This compares to only about a 40 percent homology between the two strains of the human immunodeficiency virus, HIV-1 and HIV-2.

Essex estimates that the SIVs have been in African monkeys for a long time—"maybe for thousands of years," he says, "but less than hundreds of thousands, or millions, of years." He knows they cannot be ancient infections because they are not found in the wild in Asian monkeys at all. If SIV had been in Africa millions of years ago, *before* the Asian monkeys split off and migrated to a new continent, the Asian monkeys would have taken SIV with them, and infection would be worldwide. But because of the global distribution of SIV—that is, its restriction to Africa—scientists can estimate that it emerged in Africa *after* the Asian species had separated.

This creates an interesting natural experiment, along the lines of the natural experiment that occurred with the Australian rabbits and the Brazilian myxoma virus. The African monkeys have evolved a peaceful coexistence with SIV—just as the rabbits of Brazil evolved a peaceful coexistence with myxoma—that allows them to carry around chronic infections without developing disease. The Asian monkeys have had no such exposure, and therefore they lack tolerance to African SIV. But when Asian macaques are housed alongside African sooty mangabeys or green monkeys in primate research facilities in the United States, they are artificially exposed to SIV, in much the same way that the Australian rabbits were artificially exposed to the myxoma virus.

Of the Asian macaques inadvertently infected with SIV, all of them got sick and died. (The same thing happened with the first wave of Australian rabbits infected with myxoma virus.) Animals never before exposed to a particular virus die off quickly, because all the animals in the population are vulnerable to the virus's lethal effects. But after many generations—perhaps as many as ten or fifteen—the animals remaining are those that have some inborn natural resistance, which they pass on to their descendants. And the remaining viruses are those that were a little less lethal to begin with, since all the highly virulent ones died off with their original hosts. Eventually, just as the Australian rabbits did with myxoma, and just as the African sooty mangabeys did with SIV, the Asian macaque might be expected to achieve a kind of equilibrium with the virus that was at first so devastating.

How did those captive macaques get infected with SIV in the first place? Some scientists believe they were exposed simply by being caged in proximity to the natural carriers from Africa. But Max Essex doesn't think that's what happened. SIV, like HIV, can be transmitted only by intimate contact and exchange of blood or body fluid; in most primate facilities in the United States, he says, different species are housed so far apart that sexual contact or biting and scratching would be all but impossible. "More likely is that laboratory primate experts inoculated the macaques with those viruses fifteen or twenty years ago," he says. "They thought they were only inoculating them with malaria or some other disease they were studying, because retroviruses hadn't even been discovered at that time."

Despite their close genetic similarity, HIV-2 seems to cause far fewer symptoms than HIV-1. People infected with it show fewer severe immune system irregularities, develop fewer opportunistic infections, and are more likely to live a normal life span and die of some unrelated disease. Some scientists explain this by saying that HIV-2 has a latency period of about twenty years, very long compared to HIV-1's ten. Others say it occurs for the same reason that SIV seems to cause no symptoms in African monkeys: because it is a low-virulence strain to begin with, or because over time it has evolved to a lower virulence. (This last possibility is unlikely, Essex argues, since HIV-2 has existed in humans for no more than one hundred years, probably not long enough to have changed either the virulence of the agent or the resistance of the host.)

The SIV story may be a sort of fugue, invoking a theme that will ultimately be repeated on a grander scale in human beings. Just as African monkeys eventually learn to live with SIV, so have human beings learned to live with HIV-2. But just as Asian monkeys are killed quickly by the same SIV that causes no harm in Africa, it may be that human beings living outside the remote regions of western Africa, where HIV-2 is endemic, are suffering the first wave of a virus that has crossed species lines. That virus might be what we now know as HIV-1.

This scenario leaves unanswered one essential question: What happened, earlier in this century, to cause SIV to infect a human being?

Because of the close genetic relationship between SIV and HIV-2, it is quite possible that the monkey virus did not have to mutate at all to turn into a human virus. But what *was* required was a change in the relationship between monkey and man, just as happened in the primate research facilities to allow SIV to leap from sooty mangabey to macaque.

It seems a straightforward question: How did SIV get into people? But nothing about AIDS is straightforward, least of all speculation about who or what in Africa set the process going. The history of infectious disease is peppered with many instances of microbes that cross species boundaries. But in no other case have scientists speculated that those leaps were facilitated by cross-species sexual intercourse or other unnatural rituals; simple geographic proximity was always considered to be enough of an explanation. In the case of AIDS, though—no doubt because of its associations with homosexual sex, prostitution, and drug abuse—a huge cesspool of innuendo, stereotypes, and xenophobia is stirred up by the very asking of such a question. And even cool-headed scientists occasionally take a dip into those murky waters, coming up with some rather outlandish theories for how the AIDS virus made its cross-species leap. Among them are the assertions that:

- SIV got into humans as a contaminant of a vaccine, especially polio vaccine, that had been prepared using SIV-infected monkey cell cultures.
- Blood inadvertently contaminated with SIV was used in the Zairian tribal ritual of injecting monkey serum into the pelvic region of both men and women (he-monkey serum for men, she-monkey for women) to heighten sexual arousal.
- HIV arose after a laboratory manipulation of SIV, concocted as an agent for germ warfare and accidentally—or maybe even deliberately—released into the environment.
- SIV was passed from monkey to man during an act of interspecies sexual intercourse, which is common among African tribesmen.

The truth is probably far more mundane. Many tribes in Africa hunt monkeys and butcher them for their meat; others actually keep monkeys as pets, the way Westerners keep dogs or cats. In any of these activities, it's quite easy to imagine a human's getting cut by a hunting knife or meat cleaver, or being scratched or bitten by a pet and receiving a good healthy dose of monkey blood through the wound opening.

Bizarre as they are, theories about interspecies sex would only serve to explain the existence of HIV-2, the rarer form of the AIDS virus. It would not explain where HIV-1 came from, which is so different from any known SIV that most American and European scientists believe it never was a monkey virus. Instead, most scientists now believe HIV-1 has existed as a chimpanzee virus, and maybe even as a human virus, in central Africa for many generations.

Richard Krause and several of his colleagues at the National Institutes of Health have developed a scenario for the origin of AIDS that places all of the early action in Africa. It's a rather controversial point of view—except among the leaders of the scientific establishment, like Krause—because it seems to work so hard at placing the "blame" for AIDS on countries that are helpless to defend themselves. The farther removed the United States can believe itself to be from the origin of AIDS, critics of this theory say, the easier it is for Americans to approach the disease as a plague visited upon them rather than as a mess of their own making. Because scientific theories about AIDS get so entangled in politics, it becomes difficult to separate the jingoism from the truth. This is not to say the theory that AIDS originated in Africa is necessarily wrong, just because it also happens to be politically expedient. Representing as it does the predominant thinking now of scientists all around the world—including not only Krause but also Max Essex, his Harvard colleagues William Haseltine and Ronald Desrosiers, and the co-discoverers of HIV-1, Robert Gallo of the United States and Luc Montagnier of France—the belief that AIDS began in Africa calls into play many of the elements of viral emergence with which we are becoming familiar: behavioral change, movement of people into cities and across oceans, and possibly slight changes in the virus itself.

HIV-1, according to this version, probably existed for many years in some remote sections of central Africa. Over the years, the virus and its human hosts evolved a relatively peaceable coexistence. The virus spread only sluggishly, because it was so difficult to transmit, and was confined to a few isolated rural populations. And although the virus most likely caused the same immune system abnormalities we now recognize as AIDS, these symptoms went unnoticed on a continent in which it was not the least bit surprising to see young people dying of all manner of strange infections.

Behavioral changes in the 1970s made the transmission routes that the virus used—the exchange of semen or blood—much more common. Colonial forces were leaving Africa; democracy was taking hold across the continent. Along with independence came urbanization, the breakup of traditional families, more sexual activity, and the introduction of Western needles for medical and ritual use (and reuse). This meant that for the first time, HIV had a whole new set of pathways for getting around. It took only a few international travelers to allow HIV to, as the epidemiologists put it, "go global."

Consider, for example, the nation of Zaire, thought by many to be the probable point of origin of HIV-1. First colonized as the Belgian Congo, Zaire (at the time named the Democratic Republic of the Congo) split from Belgium in 1960. Traditional values were generally discarded by Zairians moving in great waves to the capital city of Kinshasa (formerly Léopoldville). At the time of independence, Léopoldville had already swelled from 10,000 people to 400,000; after independence, the population erupted, reaching 2.5 million in 1980 and perhaps 4 million today. With this surge of people came squalor, pestilence, poor sanitation, rampant infectious disease. And with the lifting of strict village social constraints, the new city dwellers engaged in widespread sexual promiscuity, including male and female prostitution. The result was an increase in venereal diseases, most of which were left untreated—and all of which are known to increase the "take rate" of HIV, improving the likelihood that a single exposure to the virus will turn into a full-blown case of AIDS.

At about the same time, hypodermic needles came to Africa. For the aura of scientific respectability, medicine men began delivering their traditional compounds with Western needles; to economize, they used the same needles again and again. Some observers also noticed the coincident arrival on the continent of international public health volunteers armed with smallpox vaccines. For Africans who might have harbored a subclinical HIV-1 infection, some say, inoculation with a large dose of vaccinia virus might have created a co-infection that activated the HIV. No good evidence for this theory exists, and the World Health Organization, for obvious reasons, denies any connection between the smallpox eradication campaign it coordinated and the subsequent AIDS epidemic. But in at least one instance, an American soldier who was HIV-positive but generally well in 1987 suddenly developed fulminant AIDS soon after being vaccinated for smallpox (which was still being done at the time by the armed forces); he died within weeks.

After the Belgians left Zaire, the Haitian professionals arrived. Because they were well educated, black, and French-speaking, Haitians were welcomed into positions of authority to help get the fledgling nation on the right track. Haitians became the architects, engineers, physicians, economists, and other professionals who replaced the colonials. But in a surge of nationalism in the mid-1970s, the Zairians expelled the Haitians. Those who returned to their homeland confronted grinding poverty, even worse than the poverty they had seen on the streets of Kinshasa. In Port-au-Prince, the cycle of disease began again: overcrowding, filth, prostitution, the quick transmission of infectious and venereal disease. And from Haiti, there was soon a direct link to the United States and western Europe. Gay men found Haiti an attractive and convenient vacation spot, and many of them sought the services of young Haitian boys who were engaged as prostitutes. Soon the sexually transmitted HIV, which Haitians had brought out of Africa, made its way into the bloodstream of Europeans and Americans. These men took HIV home with them and continued to spread it among the men they met in the gay bars, bathhouses, and parties of San Francisco, Los Angeles, New York, and Paris. The AIDS pandemic had begun.

The problem with this nice, neat chronology is that it fails to take into account some very old case histories that with hindsight look a lot like proto-AIDS. As far back as 1868, for instance, the Viennese physician Moriz Kaposi saw the first of several strange cases of a sarcoma that would later bear his name—and is now recognized as a hallmark "opportunistic infection" of AIDS among young gay men. Five sarcoma patients consulted Kaposi between 1868 and 1871; in many respects these men, who died 125 years ago, sound as though they had AIDS. The autopsy of one patient, for instance, revealed not only the lesions of Kaposi's sarcoma throughout his internal organs but also unidentified lesions in the lungs that might have been the *Pneumocystis carinii* pneumonia that kills so many AIDS patients today. Another mini-epidemic of Kaposi's sarcoma occurred just a few years later, between 1874 and 1882, among twelve patients in Naples—one five-year-old child and eleven men between the ages of thirty-nine and forty-four. Although physicians then never recorded the personal histories or sexual orientations of their patients, both Vienna and Naples were known during the late nineteenth century to be active centers of homosexuality.

Again and again, mysterious, unprecedented infectious diseases were recorded in medical journals and left unsolved. The reports were too scanty for even a retrospective diagnosis, because of the injunction against giving away too many personal details in these case studies. But many of those infections were found by pathologists at the time to be caused by organisms we now recognize as "opportunistic." And occasionally, an astute observer kept a tissue or blood sample for posterity, knowing that pathologists of the future might have a better idea of what caused a particular patient's death. In those few cases, the mysteries have indeed subsequently been solved. And the findings have pointed to confirmed cases of HIV infection that date back as far as 1958.

This earliest case involved an English sailor, aged twenty-five, who began in late 1958 to show symptoms of gum inflammation, skin lesions, breathlessness, fatigue, weight loss, night sweats, cough, and hemorrhoids. By February 1959 he was hospitalized for a large painful fistula (an abnormal tubelike cavity) in his anus and a small tumor in his nostril; both continued to grow, eventually becoming ulcerated. Sir Robert Platt, president of the Royal College of Physicians, was called in as a consultant; all he could guess was that the sailor was suffering from an unknown viral disease. What he wrote in the man's chart turned out to be highly prescient: he wondered "if we are in for a new wave of virus disease now that the bacterial illnesses are so nearly conquered."

The sailor died in September 1959 at the Royal Infirmary in Manchester, having failed to respond to antibiotics, steroids, and radiation therapy. George Williams, the hospital's pathologist, performed an autopsy and found an unusual co-infection with cytomegalovirus and *Pneumocystis carinii* pneumonia. He thought the case curious enough that he wrote it up in *The Lancet* as a medical mystery, and preserved samples of the patient's tissue in paraffin.

Twenty-six years later, in 1985, Williams and his colleagues tried

to run tests on the patient's preserved tissue to see if the newly discovered AIDS virus could be detected there. The sailor's symptoms,
after all, seemed like a classic case of AIDS—though no one knew
anything about his sexual orientation, and his relatives could not be
found. But if there was HIV in the tissue, it was in too small a quantity
to be detected.

All that changed with the commercialization in 1988 of a new test
called polymerase chain reaction (PCR). With PCR, sequences of
genetic material that exist in extremely low concentrations can be mul-
tiplied into enough identical sequences to be detectable by conventional laboratory methods. The multiplication occurs at dizzying rates
of speed. If a tissue sample has just a single virus particle, PCR can
turn that into *ten billion* virus particles in a single hour. Virologists like
Stephen Morse have had great fun imagining and actually trying out
the ways in which PCR can help answer some of the questions that
most intrigue them. Morse, for one, sees its greatest value as a tool in
"disease archaeology," such as has been applied to early AIDS. The
test gets around the biggest difficulty with conventional virologic
methods: that it is so easy to disrupt a virus sample. "It is often difficult
to detect viable virus in fixed tissues or in samples that have been stored
carelessly or for long periods," Morse says. "But by PCR, many otherwise intractable samples can now be tested, maybe even mummified
human bodies several thousand years old."

That is just what happened with the sample of the British sailor's
tissue, which, by the time it was resurrected from its suspended animation, had been in storage for thirty-one years. In June 1990, after
eight months of experimentation in which PCR played a critical role,
George Williams and his colleagues had their answer. They successfully cultured tissue from the sailor's kidney, bone marrow, spleen,
and the lining of his mouth. To check their results, they also cultured
stored tissue from a young man who had died in a car accident the
same year the sailor died. None of the accident victim's tissue cultures
had the gene sequences of HIV. But every one of the sailor's tissue
cultures did.

Another very early European case of AIDS also occurred in a
sailor, this one a Norwegian. In his case, he also transmitted the disease
to at least two other people: his wife and their little girl. When the
father was a sailor in the mid-1960s, he traveled widely throughout
Europe and Africa. In 1966, at the age of twenty, he went to his phy-

sician complaining of swollen lymph glands, muscle pain, recurrent colds, and dark spots on his skin. His disease flared up again nine years later—by which time the sailor had married and fathered three daughters—and he died with lung and brain disturbances in April 1976. Four months earlier, his youngest daughter, aged nine, died of disseminated chickenpox after having suffered from a series of severe infections from the age of two. And later that year, in December 1976, his wife died too, of acute leukemia and encephalitis. The two older girls remained perfectly healthy.

The Norwegian's disease was so mysterious, and its ravaging of his unfortunate family so disturbing, that once again samples were preserved: his physicians froze his blood and that of his wife and daughter in the hopes that some day an explanation could be found. In 1988, it was finally time to analyze the blood samples. Every one of them was shown to have antibodies to HIV.

In the United States, the first confirmed case of AIDS dates back to 1968. The case involved Robert R., a fifteen-year-old black American living in St. Louis. In 1968, Robert was admitted to the St. Louis City Hospital for grossly swollen genitals and legs, which would have looked like the disease known as elephantiasis except that disease usually occurred in the tropics. Robert had never been outside of the United States; indeed, in his short life he never even left St. Louis. Mystified by this unusual illness, his physicians took samples from his lymph, blood, and prostatic secretions; in every one, they isolated a sexually transmitted microbe called *Chlamydia trachomatis*. Though the infection usually responds well to antibiotics, Robert did not get better. He remained in the hospital, wasting away, failing to show an immune response to any treatment, and he died on May 15, 1969. As far as his sexual history, Robert admitted to a single encounter with a girl, but his autopsy told a different story. His body revealed signs of anal trauma usually associated with homosexual intercourse, as well as internal lesions typical of Kaposi's sarcoma.

Robert's surgeon, Marlys Witte, and her colleague Memory Elvin-Lewis thought to freeze their patient's blood and lymph samples, reasoning that at some point in the future the mystery of his death might be explained. They were right. Eighteen years after the young man's death, vials of his blood and lymph were unfrozen by Robert Garry, a microbiologist at Tulane University. All samples carried antibodies to HIV-1, as well as HIV-1 antigen.

What does it mean for these sporadic early cases of AIDS to have erupted in England, in Norway, in the United States? Where did they come from, and why didn't they spread? What happened to the virus during the ten years or more between these isolated first cases and the earliest wave of the current AIDS pandemic? One possible conclusion that can be drawn, according to Mirko Grmek, is that AIDS did not necessarily arise in Africa at all; it may be a purely American or European phenomenon. "The AIDS agent existed for a long time in the Western world," he writes, "and it was unnecessary to invoke its introduction from Africa to explain an epidemic whose wildfirelike spread depended exclusively on certain lifestyle changes." In this view, AIDS began as a few tiny sparks that kept flaring up around the world, and the social changes of the 1970s—as well as the conquest of other infectious disease that had been elbowing AIDS out of the way—were sufficient to turn a few of those self-limiting fires into a worldwide conflagration.

When this international flare-up began to take hold, at first it was good old-fashioned sex that got things under way. Prostitution in Africa, Europe, and the Americas; polygamy in Africa (as well as the tradition of male survivors' marrying their dead relatives' widows); anonymous homosexual sex in American and European cities—all contributed to the rapid spread of what was essentially a new and highly lethal venereal disease. But there was a twist with AIDS. Because it was carried in the blood, and because modern medicine had concocted especially ingenious ways to spread blood around, HIV was also being transmitted—at least in the years before the virus was identified—by purely iatrogenic means.

Early in the epidemic, without a way to check donated blood for the presence of antibodies to HIV, AIDS was frequently transmitted through contaminated blood during emergency transfusions. Claude Chardon was among the first such cases after his 1978 blood transfusion in Haiti; but by 1983, the AIDS virus was becoming a real problem in the American blood supply as well. From 1983 through 1986, an estimated 1.5 to 2.5 percent of American cases of AIDS, and 10 percent of AIDS in American women, could be traced to receipt of a contaminated unit of blood during an emergency blood transfusion. As Grmek put it, "An act of mercy, a triumph of modern medicine, had become a mortal menace."

Medical innovations also led to another way to deliver even more potent doses of blood—and, unwittingly, of the AIDS virus too—by pooling blood donations to make particular biological products. The most prominent of these was in the manufactured blood components that in the past twenty years have become a lifeline for hemophiliacs. Boys and men with hemophilia (it never affects females) lack an important component in the blood that allows the blood to clot as a way to stanch bleeding. If they receive regular injections of the blood-clotting component they lack—usually a shot every one to four weeks—they do fine; if they don't, a minor injury could make them bleed to death. This component, known as factor VIII (or, in a rarer form of hemophilia, factor IX), is made from donated blood from which the blood-clotting factor has been purified and concentrated. A single lot of factor VIII comes from the pooled blood of, on average, twenty thousand donors. So any one hemophiliac is exposed to the blood of literally millions of donors over the course of his lifetime. Before 1985, when blood banks began screening donated blood for HIV antibodies, hemophiliacs ran an astronomical risk of encountering HIV in at least one of their many doses of factor VIII. Of the nation's twenty thousand hemophiliacs, one-half became infected with the AIDS virus by 1985. That fraction may be even higher among those with severe hemophilia, since they need more frequent doses of factor VIII.

Modern medicine etched another pathway for HIV through the use of organ transplants, ironically snatching life away from the very individuals whose lives had been spared by high-technology surgery. With mounting horror, epidemiologists at the Centers for Disease Control have recorded this grim side effect of medical heroism—the introduction of a lethal virus along with a life-sustaining organ. Although this has been a rare occurrence, in one dramatic instance a single organ donor actually infected at least four people, and the toll could continue to rise. The young man involved was William Norwood of Richmond, Virginia, who was shot to death in 1985 during a robbery at the gas station where he worked. Norwood, who was twenty-two at the time of his murder, had been registered as an organ donor, and after his death fifty-four tissue grafts were taken from his body and implanted in at least fifty individuals. The company responsible for processing Norwood's tissues said they had tested his organs twice for HIV antibodies, and both times the results were negative. It was not until six years later that anyone discovered that Norwood had in fact been infected with HIV but died before his body mounted an immune response. As a result of those transplants, as of mid-1991, three transplant recipients were dead of AIDS, and one more tested positive for HIV.

The question of where AIDS came from, and where it may be going, surfaced again in 1992 at the Eighth International AIDS Conference, held that year in Amsterdam. Spurred by their colleagues' informal comments in hotel hallways and seminar rooms, scientists collected case reports of a few dozen people scattered around the world who seemed to have AIDS without any signs of HIV-1 or HIV-2 infection. These patients had no HIV antibodies and no trace of HIV even when examined with the highly sensitive PCR test. Then one scientist made a hastily scheduled presentation about his immunosuppressed patient, a sixty-six-year-old woman with PCP, and her apparently healthy thirty-eight-year-old daughter. Neither had any HIV in her blood—but both showed signs of infection by a new retrovirus never before associated with human disease.

Criticized for not having acted earlier on these perplexing developments, CDC officials convened a meeting of leading virologists as soon as everyone returned from the Netherlands. The consensus: no one knew exactly what was transpiring. Had this AIDS-like illness been occurring sporadically for years, only to be uncovered now because of increased surveillance? Was the new retrovirus a cause of the condition or just a coincidence? Were all these problems caused by a single new agent? And did that new agent threaten the safety of the nation's pool of donated blood?

If nothing else, the questions raised by these thirty or so cases made one point disturbingly clear: nothing about viruses, especially a virus like HIV, stays static for long. And that means that nothing about viruses is easy to investigate. Looking at blood samples might reveal "footprints of virus," said Anthony Fauci, director of the National Institute of Allergy and Infectious Diseases, "but all of us who've been in the laboratory for any length of time have been down that road before—where you're thinking you've found something, and in fact it's either nothing or it's unrelated to what you're looking at."

In many ways, the very presence of HIV, in the clinic and in the laboratory, creates its own problem. Consider, first, the increased longevity of AIDS patients, now that early treatment with the antiviral drug AZT (azidothymidine) allows people to live for five or ten years despite their HIV infection. As more immunosuppressed people survive for longer periods, the odds increase that one of the many viruses they harbor in their own bodies can become reactivated and turn "opportunistic." This could create problems not only for the AIDS patient in question, but for the larger community as well.

The reactivation of some latent childhood infections, as can happen in patients receiving immunosuppressing drugs, could just as easily happen in virally immunosuppressed patients—in other words, in people with AIDS. In the new environment of an immunosuppressed adult, the latent virus may grow and change, becoming more capable of infecting other adults with functioning immune systems. What was once a harmless virus, restricted to children, could become a new danger.

A similar process could occur in the laboratory, where viruses can mutate and thrive in the unique environment of the culture dish or the laboratory animal. This scenario has been proposed by Robert Gallo, one of the most famous AIDS researchers in the United States, who worries that when scientists infect experimental mice with HIV—as has been done successfully in several laboratories since about 1988—the virus could mutate in dangerous ways. In certain strains of mice that have been genetically altered to accept human immune system cells, such mutations already have occurred. Gallo and his colleagues reported that a common endogenous (inborn) mouse retrovirus did in fact interact with HIV-1 when they were both placed into a culture dish containing human cells. The HIV-1 acquired the ability to reproduce more rapidly than normal and infect cells other than the immune system cells it usually infects—such as the cells lining the respiratory tract. Although so far the recombination has been noticed only in cultured cells, most virologists believe it is quite likely to occur in HIV-1–infected mice as well. This could mean that in the new environment of a mouse, HIV-1 could recombine and evolve in a way that allows it to spread even more quickly, and possibly even to spread through the air.

This would obviously present enormous safety concerns, were it not for the fact that even if such changes did occur in the lab, the animals are ordinarily kept in the very highest containment conditions available. This does not always work, of course: there may be a break in safety technique or an unavoidable accident, such as a bite or scratch from an infected animal. But the problem goes beyond issues of safety to issues of science. If HIV-1 studied in a mouse recombines so easily with mouse retroviruses, what are the investigators really studying? Certainly not HIV-1 as it would appear in a human being. "The results in the mice can't be interpreted as easily after this," Gallo says.

The strides made in understanding HIV—not only where it came from, but, more important, how it works and how it can be stopped—could not have taken place ten years or even five years earlier than it did. AIDS came to international prominence at a remarkable juncture in the history of science, a time when molecular biology was just coming into its own, and when an understanding of immunology was at last being refined down to the level of the gene. The very existence of the category known as retroviruses was not even hypothesized until just a few years before HIV, the most devastating retrovirus known to man, appeared on the scene. Many of the more recent developments in virology have occurred because of the spotlight of AIDS research; the more HIV has been studied, the more scientists have uncovered about viruses and immunology in general. Similarly, the more that can be discerned about why AIDS emerged, the better able scientists will be to analyze where other emerging viruses might have come from—and how different ones might emerge in the years ahead. It is important at this point, though, to take a brief break from predicting and develop an understanding of some basic principles of virology. Without this understanding, the threats and the promise of emerging viruses would be all but impossible to discern.

# 3

# *A Virus Primer*

Viruses are so weird and wily that they fairly cry out for metaphors when you write about them. They have been glamorized as "pirates of the cell" and "submicroscopic hijackers"; they have been dismissed as "a piece of bad news wrapped up in protein." These metaphors are attempts to describe the essential nature of the virus: its ability to sabotage a cell's functioning, so that the cell starts churning out more virus instead of whatever else it's supposed to be producing.

Colorful as they are, though, metaphors sometimes serve to obscure rather than clarify our understanding of exactly how viruses behave. These are just microbes, after all, not swashbuckling pirates with their own anthropomorphic goals in mind. When they get inside the cell and turn on the cell's machinery for copying genes, they are simply following the instructions of their own genetic blueprint. There's nothing sinister or premeditated about it.

Still, there is one metaphor that offers a charming beginning for a quick introduction to virology. It's a restrained metaphor, one that manages to describe not only the way a virus functions but also its origins and its relationship to us, its beleaguered hosts. Peter Radetsky, a science writer from Los Angeles, offers the image of viruses as "minute, wayward, and unruly parts of ourselves—something like adventurous teenagers who have fled the nest but just can't resist coming back home at every opportunity." Radetsky carries his metaphor one step further, which is just far enough. When the viruses-as-adolescents come "back home," he writes, it's "sometimes to overstay their welcome, sometimes to wreak absolute havoc, sometimes to make us better through their mere presence. And, like loving parents, for better or for worse, we almost always leave them a key to the front door."

In many ways, viruses really are like teenagers run amok. They "wreak absolute havoc" on our bodies, they occasionally switch into a latent infectious mode that enables them to "overstay their welcome," and their "mere presence" can indeed help us instead of hurt us. And viruses are, in a manner of speaking, our own offspring—if one subscribes to a prominent theory that holds that they actually began eons ago as pieces of our very own genes.

But in the traditional understanding of the term, viruses are not really even "alive." Unlike virtually all other life-forms, a virus contains only one type of genetic material, either DNA or RNA. It never contains both. And it cannot exist on its own; outside of a compatible host cell, a virus is as inert as a piece of paper. Without the host, a virus cannot reproduce, cannot metabolize, cannot conduct any of the functions that identify a thing as being alive.

Every single virus of a given type is exactly the same size and shape as every other virus of that type. No other living thing has this quality; everything else that is alive, from oak tree to elephant, begins small and grows. And most groups of viruses are capable of forming highly geometrical arrangements known as crystals, a behavior more typical of chemicals than of cells.

A virus behaves differently from any other pathogen, or disease-causing agent. Almost every other pathogen, except for the strange agent known as the prion, is a single-cell organism, with its own cell membrane, its own genetic material, and—with one prominent exception—its own cell nucleus, the cellular compartment that houses the cell's genetic material in every organism on earth, from protozoan to human being. The exception is the bacterium, which is too primitive to have a nucleus. Its genetic material is not concentrated anywhere; bacterial genes simply float in the cell cytoplasm, the semiliquid material that makes up the bulk of any cell, and collect in a specific arrangement whenever the bacterium multiplies by splitting in two.

But a virus is not a cell. It has no nucleus, no membrane, no cytoplasm. A virus is just genetic material surrounded by one or more coats of protein. Occasionally, the protein coat is encased in an extra protective envelope made of sugars and fats. Viruses hang around the atmosphere, cocooned inside those protein coats, waiting for a way to get inside a host animal or plant. (Yes, there are plant viruses, but they never infect animals, just as animal viruses never infect plants.) They exist on dust particles, in the tiny airborne droplets of a sneeze, on an animal's fur; they lie in wait in the soil or on a tabletop, or they travel in the guts of a vector (carrier) insect. They vary in how long they can wait outside a host cell before drying up entirely, but wait they most certainly do. They are not lifeless, really, but neither are they functionally alive. They are suspended in a state of expectation, waiting to be activated.

Activation occurs once the virus gets inside a host cell. Now this is life for certain—life with a vengeance. The virus takes over the cell's reproductive machinery, forcing the cell to produce more viruses through the same mechanisms it uses to copy its own cellular genes. Thus sabotaged, the cell not only fails to perform its intended function but also helps its invader multiply. It's because of this trickery that the metaphors about viral "wiliness" are so prominent.

One of the most obvious differences between viruses and other pathogens is their size. The average bacterium is about one micron (one twenty-five-thousandth of an inch) across; the biggest virus is just one-quarter of a micron wide, and many are as small as seventeen *milli*-microns (one one-thousandth of a micron). An average-size virus such as adenovirus is so small that five billion of them—more than the number of people on earth—could easily fit into a single drop of blood.

Because they are so tiny, viruses could not even be seen until 1937, when the electron microscope was invented. Before then, their very existence could only be inferred. Martinus Beijerinck, a botanist at the Delft Polytechnic School in Holland, was the first scientist to state that infectious agents existed that were smaller than any previously identified bacterium—and that they were impossible to see. In 1898, Beijerinck transmitted a leaf-mottling disease from one tobacco plant to another with a fluid he knew could not contain any known pathogen. He ground up infected tobacco leaves into a pulpy sap and passed the fluid through a porcelain filter with pores so small they screened out all bacteria. But something in the fluid, something so small it could pass through the tiniest of filters, was still able to transmit the condition known as tobacco mosaic disease. Beijerinck called his liquid a "contagious living fluid."

The phrase was controversial, because in 1898 the belief that *anything* was contagious—not to mention the belief that the contagious particle was invisible—was just this side of heresy. The great microbiologist Louis Pasteur had presented his "germ theory of disease" less than thirty years earlier, and even at the turn of the century you could find scientists who weren't convinced that disease was caused by microorganisms, even those that could be seen. But Beijerinck was sure of his findings. And within a few years, he was saying that inside this "contagious living fluid" was an invisible microbe called a "filterable virus." In Latin, *virus* means "poisonous slime."

When the electron microscope—capable of spotting cellular components and other objects four hundred times smaller than those visible with the traditional light microscope—was introduced, viruses finally could be seen. At once, their architecture became the subject of intense fascination. Some viruses were shaped like icosahedrons, spheres composed of many small triangles, like soccer balls. Some were perfectly round, some perfectly conical. Some viruses, like influenza virus, had projectiles on every available surface, like an orange studded with cloves at Christmastime. Others, like coronavirus, were surrounded by a halo of soft outcroppings that made them look like the shining sun. A virus that infects bacteria, called a bacteriophage—one of the first to be viewed under the electron microscope—had legs and a tail in an odd cross between an upright insect and a lunar landing module. Beginning in the 1950s, when electron microscopy became commonplace in microbiology laboratories, virologists spent a good deal of their time looking at and describing the incredible array of shapes that were possible—a far more impressive array than those they had grown accustomed to after looking at bacteria or other one-cell organisms. The peculiar nature of viral diseases, it seemed, was mirrored in those bizarre, apparently lifeless shapes.

The kaleidoscope of virus shapes gives a clue to another important difference between a virus and any other life-form: its life cycle. Most of the other microbes that cause disease can live and reproduce independently; even the category of infectious agents known as parasites can manage some of the functions of living on their own. The most dangerous parasite known to man, the *Plasmodium* protozoan that causes malaria, can still eat and metabolize outside of a host. It is only to perform the function of reproduction that it must take up residence in the gut of a mosquito or in the red blood cell of a human being.

Viruses, in contrast, are total parasites. They cannot do anything on their own: they cannot eat, cannot grow, cannot reproduce, cannot respire, cannot really even move. Out in the atmosphere, this quality translates into either extreme hardiness or extreme fragility: hardiness because viruses must wait a long time for the right host to come along, fragility because they might die waiting. The smallpox virus, for instance, is as invincible as a boulder, able to "live" in suspended animation for decades and perhaps even centuries. But the AIDS virus can be as evanescent as a soap bubble. Once outside of the blood cells in which it hides, it dries up and disappears after just minutes in the air.

All viruses, be they fragile or hardy, must accomplish a few basic steps to complete their life cycles. They must penetrate a host cell, shed their protective protein coats to expose their viral genes to the cell, utilize the cell's gene-copying machinery to make more viruses, put on a new protein coat, and escape from the host cell to infect another.

To get inside the cell, a virus does one of two things: either it plunges through the cell membrane by brute force, causing a chemical change in the membrane that actually splits it open, or it fuses its own membrane with that of the cell, sliding its way inside as the membranes turn each other inside out. An interim step in this process, one used by many viruses, is a technique called endocytosis, in which the cell embraces the virus with a wide sweep of its membrane and pulls it inside, closing off the membrane behind it. Endocytosis of viruses happens almost by accident; the true targets of endocytosis are extracellular molecules that the cell must incorporate in order to maintain its functioning. To get swept in by this general process of enfolding, viruses have developed ways to look just like those molecules the cell needs to embrace. The rhinovirus that causes the common cold, for example, has portions of its outer membrane that are identical to those of molecules that are useful to the cell. In the metaphor that began this chapter, the endocytosis of camouflaged viruses is the way our own cells sometimes leave viruses "a key to the front door."

Once the virus is inside the cell, its protein coat is dissolved, either by enzymes it carries as part of its own package or—for the simplest of all viruses, like the parvovirus—by enzymes already in the cell itself. When the protective coat is gone, the cell is exposed to the genes of the virus. The viral genes direct the cell to start manufacturing more viral genes, then to follow their instructions to make viral proteins, and, ultimately, to assemble them into a string of new viruses for release into the bloodstream. In some especially pernicious infections, such as poliovirus, the viral instructions go one step further. They actually force the cell to make suicide proteins—enzymes that will actively force the cell to stop manufacturing the proteins it needs for its own survival.

Generalizations can be made about how viruses get into cells, disrupt cell functioning, and cause disease. But what is especially interesting about viruses are not the generalizations but the differences, differences that allow virologists to categorize viruses according to their shapes, their life cycles, the types of disease they cause, or—in a categorization especially enlightening for the study of emerging viruses—whether the virus carries its genetic payload in the form of DNA or RNA.

DNA (deoxyribonucleic acid) is, of course, the genetic blueprint for all forms of life. Everything alive—from bacteria to orangutans, from day lilies to human beings—uses DNA to direct all its functioning. And everything alive uses a variant of DNA called RNA (ribonucleic acid) to help make more of the right kind of DNA. But viruses are composed of either one or the other, allowing them to be categorized into two types—DNA viruses and RNA viruses. DNA viruses, such as variola (which causes smallpox) and the herpesviruses (which include genital herpes, cytomegalovirus, and Epstein-Barr virus), tend to be larger and more stable. Stability in this context means that their genetic material stays the same from one generation to the next; errors or mutations in copying the genes of DNA viruses do occur, but they are relatively rare. RNA viruses are one hundred times as likely to mutate as are DNA viruses, because they can short-circuit the highly regimented procedure of genetic reproduction. They miss an important "editing" step in the replication process that ordinarily helps eliminate copying errors.

Because their mutation rate is higher, RNA viruses are more flexible than DNA viruses. Take influenza, a common RNA virus. Within any group of influenza viruses are large numbers of mutants. Most of them are eliminated by natural selection because they are less able than the predominant strain of influenza to survive, which reflects centuries of evolutionary adaptation. But if the environment changes, the survival advantage of the older strains might disappear. In a new environment, some of the mutants might turn out to have the advantage. This genetic flexibility allows the influenza virus to take advantage of new environmental niches, to move to new geographic locations, to cross species lines—in short, to become what we are calling an "emerging" virus. Among the emerging viruses are many other RNA viruses: Hantaan virus, Borna virus, the arboviruses (those carried by insects, such as dengue fever, yellow fever, and various encephalitis viruses), and the arenaviruses (which are carried by rodents, like Lassa virus).

An important subset of RNA viruses, the retroviruses, use a particular bit of sleight-of-hand to force the host cell to make more of them. Until ten years ago, scientists thought retroviruses were of veterinary interest only; they thought there were no retroviruses that infected humans. But the emergence of HIV, the retrovirus that causes AIDS, has changed that way of thinking. (Actually, HIV was the second human retrovirus to be identified.) At last count, there were five known human retroviruses. And they cause great damage, usually decades after the initial infection: one causes AIDS, another causes a type of cancer called T-cell lymphoma, and the most recently described retrovirus is said by some to cause the puzzling condition known as chronic fatigue syndrome.

Boiled down to its elements, the cell of any living thing must complete one of two assignments: either reproduce its DNA for growth and reproduction, or manufacture protein for cellular functioning. Once it has grown to full size, a cell's main job is to produce protein, which is what gives it its unique characteristics. A kidney cell differs from an eye cell because of the proteins each one produces; a leaf cell differs from a petal cell because of protein production too.

All cells of an organism contain exactly the same DNA in exactly the same arrangement. In every cell of a human body, for example, there are exactly the same one hundred thousand or so genes. All human cells are capable of making any human protein, but only certain proteins are manufactured by any one type of cell; and only the genes that direct the manufacture of those proteins are switched on, or expressed, in that particular cell. So in a kidney cell, the regions of DNA that make kidney proteins—those needed to produce kidney enzymes, specialized membranes, filtration systems, and the like—are in the On mode, and all other regions are switched Off. And in an eye cell, the only regions switched on are those that make proteins needed for eye functioning.

To understand how viruses disrupt this process, we first need to understand how DNA is copied, and how it directs the production of proteins, in a cell that is *not* infected. The procedure of copying one string of DNA to make two is known as replication. The process of converting a single strand of DNA into a single strand of RNA is known as transcription. DNA is transcribed so that the RNA can be used as a template, a mold from which the relevant protein can be constructed. The process of protein-building, with the end result a particular enzyme or other protein, is called gene expression.

The building blocks of DNA are subunits called nucleotides, of which there are only four. They are the same four for every form of life on earth. From these four nucleotides, any of twenty amino acids can be built, each one coded by a three-nucleotide group. And from those twenty amino acids, the cell is capable of stringing together long, folded proteins that accomplish all the functions of the organism. Because of the proteins our cells make, we are human beings and not squid.

In the shorthand of molecular biology, the nucleotides that make up DNA are known by their initials: A, T, C, and G. They are lined up along the DNA molecule like beads along a necklace.

The structure of DNA is something like a spiral staircase, with a split down the middle. On one half of each step is one nucleotide; on the other half is its complementary nucleotide, which fits into it like a jigsaw-puzzle piece. Whenever A is on one half of the stair, T is on the other half, and vice versa; whenever C is on one half of the stair, the other half is G.

As the cell gets ready to divide, the staircase splits in two. This means there are two half-stairs now, one strand that represents the "positive" strand of nucleotides lined up side by side, and the other, representing the "negative" strand, composed of its complementary nucleotides in the same order. The two strands are biological mirror images. If the positive strand of one section of the gene is A-T-A-G-A-C, the negative strand is inevitably T-A-T-C-T-G.

The negative strand, then, serves as a template for making a new positive strand—the strand that directs the cell in the manufacture of proteins. The order of the nucleotides in the negative strand imposes an order on any positive strand it makes; all new strands will contain nucleotides in exactly the reverse (or, more precisely, complementary) order. In this way, every new positive strand of DNA is exactly the same as the original.

This is what happens during DNA replication, the copying of one

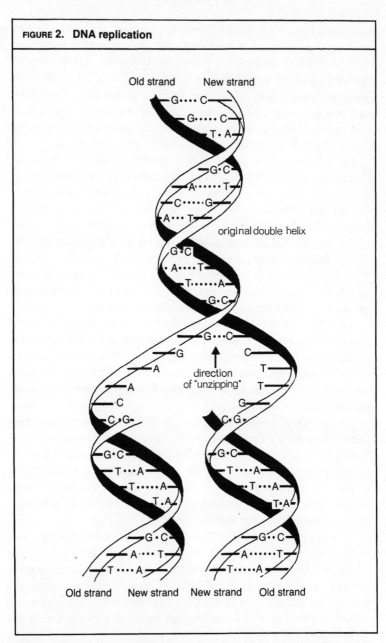

FIGURE 2. DNA replication

Figure 2. The double helix of DNA replicates by serving as its own template. The two halves of the helix separate, leaving each strand exposed. Then nucleotides, circulating in the cell cytoplasm, are brought to the "unzipped" DNA, where they bind to their complementary nucleotides. The result: two identical double helices where one had been.

DNA molecule into two. The process of protein production begins the same way—but instead of just copying the positive strand, the goal is to make a positive strand of a related molecule, RNA, which in turn directs the manufacture of new protein molecules.

RNA, the chemical cousin of DNA, is exactly the same as DNA with two exceptions: it uses a different compound along the bannister of its spiral staircase, and it uses one different letter in its four-letter alphabet, U instead of T. So when the negative-strand DNA we started with, T-A-T-C-T-G, makes RNA, it can lead to only one arrangement of nucleotides: A-U-A-G-A-C. This new RNA, called messenger RNA (mRNA), looks almost exactly like the positive-strand DNA we started with. The only difference is that there is a U in the second position instead of a T.

Now the mRNA gets down to the task of making protein. It heads out into the cytoplasm toward the ribosome, a tiny intracellular structure used for protein assembly. The ribosome reads off the triplets of nucleotides along the mRNA chain to determine which amino acid is called for. Then it directs a different class of RNA, a biological retriever called transfer RNA (tRNA), to fetch the right one. The first triplet in our mRNA example, AUA, stands for isoleucine, so tRNA hunts out isoleucine from among the stew of amino acids suspended in the cell body. Then it brings it to the right spot on the ribosome and goes off to fetch amino acid number two: in this case aspartic acid, coded as GAC. With the tRNA doing the grunt work and the ribosome doing the assembly, the mRNA chain thus gives the orders for which amino acids should be retrieved, and eventually a protein chain is formed.

A, T, C, G, and U are the entire alphabet of biology. The amino acids these letters signify can be thought of as little words; the words, in turn, produce the proteins that are the equivalent of biological sentences. Amino acids are always coded for by exactly three letters (nucleotides). Proteins can be made up of anywhere from a few hundred to a few thousand words (amino acids). And a gene comprises however many DNA nucleotides it takes to make up an entire sentence (proteins). So one gene can consist of anywhere from a few thousand nucleotides to tens of thousands, all in a row.

We fall once again into the habit of metaphor: nucleotides are letters, amino acids are words, proteins are sentences. It's uncanny, really, how well the DNA code fits into the metaphor of language—until you remember that the code itself is a human construct. The DNA blueprint, the entire description of DNA's methods of replication and expression, is all an invention. The language metaphor explains DNA functioning as well as it does because even scientists' most sophisticated understanding of genetics is itself based on an analogy to language.

At the time the genetic code was first broken, in the 1950s, linguistics was just coming into its own. "This obviously influenced our thinking," says Andrzej Konopka of the National Cancer Institute, a leader in the new subspecialty known as biolinguistics. "Most biologists were mentally ready to think of the genome as a communications system."

Scientists like to think they are conduits for some basic truth; they discover how the world works, and they tell the rest of us. But when they use human language to describe the biological language of DNA, the true nature of scientific "discovery" becomes clearer. Like all other human beings, scientists impose their own view of the world on the mechanisms they seem to have uncovered. If we were some highly intelligent species that used a method of communication quite different from written or spoken language—one that relied only on eye movements, for instance, or on verbalizations comparable to those of songbirds—maybe we would have made sense of the workings of DNA using some musical or kinetic imagery. There is nothing inherently linguistic about the way genes work; that's just the convention that verbal people have chosen to explain it.

In fact, using language to explain *any* of these biological processes is highly artificial. And, paradoxically, the most lucid-seeming explanations might sometimes make these processes harder to understand. We say the structure of DNA is like a spiral staircase split in two— but how does that help us visualize what replicating DNA really is doing? We say that ribosomes are like factories and tRNA is like the cell's retriever—but what exactly does that mean inside a living cell? These staircases are splitting, these factories are working, these retrievers are going out to fetch nucleotides *in every cell of your body, all the time*. No metaphor could approach the soul-stirring magnitude of that unimaginable fact.

The bare-bones outline of our supershort course in molecular biology is that DNA makes RNA, and RNA makes protein. Indeed, this blueprint for cellular activity was for many years believed to be the only way that life could work: in one direction only. The belief that all cell activity went from DNA to RNA to protein was so entrenched during the 1960s that Francis Crick, the co-discoverer of the structure of DNA, called it the "central dogma" of molecular biology. But some viruses have given the "central dogma" a good rattling. Many viruses, it turns out, carry their genetic information in unconventional ways—ways that often mean that RNA has to be converted to DNA instead of the other way around.

The very existence of RNA viruses seems to fly in the face of the central dogma. How can DNA always make RNA when some living things exist that have no DNA at all? But on closer inspection, it seems that most RNA viruses do not really contravene the central dogma; they just take a shortcut. Instead of beginning with DNA, they manufacture mRNA directly from their own viral RNA.

RNA viruses are categorized according to whether they are positive-stranded or negative-stranded; in other words, whether the genetic information they carry into the cell looks just like the cell's own messenger RNA or just like its complement. Positive-strand RNA viruses, which include poliovirus, yellow fever virus, dengue fever virus, and the exotic togaviruses, can and often do function exactly as if they were messenger RNA when they invade the cell; they can direct protein production just as they are. Negative-strand RNA viruses, such as influenza, rabies, measles, and certain arboviruses, must first convert their RNA into a complementary strand in order to direct the host cell to manufacture viral proteins. A few RNA viruses, such as the reovirus, are actually double-stranded, but they split in two and use their positive strand as a replacement mRNA. In all these cases, then, viral RNA is made directly into viral protein; DNA never enters the picture.

One class of RNA viruses is an exception—so much of an exception that it actually *does* fly in the face of the "central dogma." These viruses carry their genes in the form of single-strand positive RNA, but once inside the host cell they manufacture a double-strand *viral DNA* that heads right for the cell's nucleus and inserts itself into the host cell's DNA. As with other viruses, these viruses force the host cell to manufacture new viral genes as it goes through its normal processes of gene expression. But unlike other viruses, this class of virus is able to integrate its genes into the cell's own genetic makeup, spliced right into the cell's DNA. Because the viral genes have taken the form of double-strand DNA and gotten into the cell nucleus, they look more like the cell's own genes than do the genes produced by any other category of virus. So they disrupt not only the cell's protein production but its reproduction too. When the cell splits and forms a new generation, the viral genes are passed on to the offspring.

In order to have made this double-strand viral DNA, the virus began with a single strand of viral RNA. Viral RNA to viral DNA— it's a totally backward direction. That's why these viruses are known as retroviruses, from the Latin word for "backward."

Retroviruses were first inferred in the 1960s by Howard Temin, a young virologist at the University of Wisconsin. He found that the actions of an RNA virus known as Rous sarcoma virus could be inhibited by a drug that interfered with DNA activity. The only way this drug could have an effect on an RNA virus, Temin believed, was for the virus, at some time in its life cycle, to make DNA. He published his theory, which went against Crick's central dogma, and saw it quickly dismissed by his older peers. If anyone made mention of it during the 1960s, it was to deride the theory as "Teminism."

The skepticism changed in 1970, when Temin and another virologist, David Baltimore of the Massachusetts Institute of Technology, demonstrated how this backward activity could take place. Working independently, Temin and Baltimore each showed that certain RNA viruses carry into the cell the enzyme needed to transcribe RNA into DNA. Ever since then, virologists have hunted down this enzyme when they think a retrovirus has been nearby. The footprint left by retroviruses has proved handy to the disease detectives—especially when, in the early 1980s, they were trying to figure out what kind of virus caused AIDS. The presence in infected blood cells of this enzyme, known as reverse transcriptase, was a sure sign that a retrovirus was hiding in the genes of infected cells.

The key word, when it comes to retroviruses, is *hiding*. Because they are entirely integrated into the nucleus of the cell, with all traces of viral RNA and viral protein gone, retroviruses can essentially disappear into the cell's own workings. The immune system, which usually patrols the body's cells trying to detect and rout invaders, cannot spot retroviruses in this hidden mode. So in the face of a completely integrated retroviral infection, the immune system— usually so impressive in its ability to maintain the body's integrity—

is not only helpless; it doesn't even behave as though there is anything amiss.

This is in sharp contrast to the immune defense mounted against most other viruses. When a healthy immune system meets up with a typical virus, it begins a quick response that comes in two stages. "Nonspecific" immune responses go into action first, working blindly against any foreign invader the way a police barrier stops every car on the road. "Specific" immune response is customized to keep out one particular intruder, as policemen would be if they were patrolling the highways looking only for cars with Minnesota license plates. The nonspecific response works first because it is rapidly switched on; we are exposed to many microbes every day that never get close enough to a cell to start an infection, because the immune cells keep them at bay. Among these cells are macrophage cells (the word means, literally, "big eater") that try to engulf any microbe that makes its way into the bloodstream; once it surrounds the invader, the macrophage digests it. Another type of immune system cell, the natural killer cell, seeks out all host cells that are harboring foreign microbes; rather than killing just the free-floating pathogen, the natural killer cell destroys the entire infected cell.

All of this nonspecific activity can happen within hours, or at most days, of infection. Specific immune response takes a little longer to get going. That is because it requires the marshaling of T cells and B cells, the soldiers of the immune system—and getting the right soldier to fight any one particular invader might take a good bit of time. In fact, the T cells and B cells, more properly known as lymphocytes, are not even activated until the virus designed to fit perfectly into their receptors actually comes into the body and binds to the lymphocyte's surface.

T cells come in four types, each one with a whimsical name that gives a hint of its function. In chronological order—that is, from the first line of defense in an infection to the last—they are the inducer T cell, killer T cell, helper T cell, and suppressor T cell. The inducer T cells are as nonspecific in their action as are the macrophages and the natural killer cells. They circulate in the bloodstream looking for an indication that something foreign has gotten in. When an infection is under way, it is the inducer T cells that release interleukin, a special chemical that lets the other T cells know it is time to begin an attack.

Once the killer T cells are activated, things get quite specific. Every killer T cell is programmed to bind with only one virus (or other pathogen). This programming was accomplished before birth by the thymus gland, a gland in the chest that orchestrates the immune system during fetal life and infancy. The thymus gland gradually fades away, a bit like the Cheshire cat, shrinking during childhood as its activity diminishes. By puberty, the thymus has entirely disappeared. But because of the work of this short-lived gland, people are born with enough different T cells to combat at least one million different invaders. When a person comes in contact for the first time with a particular antigen (any invader cell, such as a virus, that fits a particular receptor on a particular immune system cell), whichever T cell has the shape designed to pair with that antigen can now come into its own. The antigen's presence stimulates the proliferation of its corresponding T cell; within days, a whole colony is grown. In anticipation of this antigenic stimulation, each of our inborn lymphocytes remains in the system in a primitive, preactivated form. Like a bachelor holding out for Ms. Right, the lymphocyte awaits the arrival of the single antigen that is its destiny.

Killer T cells work by doing just what the name says: they kill cells that are infected with their target virus. They find the cells in a surprisingly straightforward way. All cells, as part of their protein-making process, display on their surfaces small bits of the proteins they have made. If they are making normal cell proteins, all is well, and the marauding T cells can recognize all surface proteins as falling into the category of "self." But if the cell is infected and is churning out viral proteins, it displays some of those proteins on its surface like a little SOS signal, a clear indication to the killer T cells that something unusual is going on inside. At the same time, every killer T cell carries on *its* surface a protein that locks into only one particular virus. If the flag and the T cell match, a linkup occurs, and the cell is destroyed.

The immune response is boosted by the helper T cells, which direct blood and lymph cells to the area of infection. This increases the odds of the right killer T cell's getting to and binding with the viral protein on the cell surface. It also brings more macrophages in to swallow up free-floating viruses. Unfortunately, this rallying of forces also leads to inflammation, which in some instances creates more problems than the cell death caused by the infection itself.

When the battle between virus and T cells is over, the last type of T cell, the suppressor cell, is called into action. These cells secrete chemicals that inactivate the immune response, theoretically early enough to keep it from overwhelming the host. Suppressor cells are also important regulators of the generalized surveillance done by lymphocytes in between infections. They are crucial in helping killer T cells determine which cells are "self" and which are "nonself"—in other words, in keeping the body from mounting an immune attack on its very own cells. When this regulation fails, the result may be an autoimmune disease.

The defensive actions of the T lymphocytes are known collectively as the cell-mediated response. Another class of lymphocytes, called B cells, fights invaders in a slightly different way. Although the B cells are actually of less importance in viral infections—they have a far bigger role in fighting bacteria—their actions have come to be thought of, in the popular understanding of immunity, as *the* primary way that infections are controlled. The B cells produce antibodies. And their action, naturally enough, is known as the antibody-mediated response.

An antibody is nothing more than a protein with an idiosyncratic architecture. Each virus has a particular region on its outer coat, characterized by a unique topography and a specific arrangement of nucleotides, that is complementary to one antibody's topography and to none other's. When the right B cell produces the right antibody for a particular infection, the antibody attaches itself to the antigen, the way two Lego pieces fit together. The antibody/antigen complex that is formed in this way is a crippled invader. Easier for a macrophage to find and engulf than a free-floating virus, it has more trouble attaching to a healthy cell and getting inside. In addition, the presence in the bloodstream of these oddly shaped molecules is a signal to certain immune system chemicals, known collectively as a complement, to produce inflammation and to disrupt the membranes of both viruses and infected cells.

Antibody-mediated immunity and cell-mediated immunity are usually effective, but they are not foolproof. The chief drawback of these specific immune responses is that each takes several days to commence. And in some cases, several days is just too long. Viruses, once inside the cell, replicate far faster than lymphocytes do. Some viruses have already forced the host cell to produce one hundred copies of themselves in the first half-hour after infection. This race to beat the virus, then, is one the immune system will usually lose. But if the virus is not a lethal one, you recover from your sluggish response to a first-time viral infection. And forever after, you have at the ready an established T cell colony, and an established group of specific antibodies, to respond much more rapidly the next time an identical virus attacks. This is what is known as immunological memory.

Tricking the body into creating these valuable immune cell colonies is at the heart of the best way to prevent a viral infection in the first place: vaccination. Vaccines deliver modified forms of viruses, those that have been either killed or merely crippled (attenuated), to get the T cell proliferation and antibody formation rolling. Because the virus in the vaccine is impotent, it doesn't make you sick. But because it contains all the viral markings the immune system is capable of recognizing, it is sufficient to create an immunological memory. Should you then come in contact with the real, dangerous virus, you will have enough antibodies from the very start to keep an infection from taking hold.


We understand how they replicate; we understand how the body tries to keep them out of our cells. But what is still nagging is truly the central question about viruses: How do they make us sick?

The short answer: it's an accident. The virus is simply going about its primary business of reproducing itself. Its main goal, even when it kills a cell, is not usually destruction. Its main goal is self-perpetuation. What the virus is programmed to do is to use a cell to make more virus, and then to leave the cell and infect another. Maybe the virus uses so much cellular raw material that none is left over for the cell itself; maybe the virus splits the cell open because it has no other means of getting new virus particles into the bloodstream. When cell death happens in this way, it is a corollary to the main action, as unintended as the death of wildflowers flattened along the side of the highway when a road-paving steamroller passes by.

Indeed, if these total parasites were perfectly evolved, viruses would have ways of replicating that involve no damage to their hosts at all. Animals that are prostrate with illness are in no position to infect other animals—and what a virus needs most is a means of transmission. And animal hosts that die, obviously, are complete dead ends for the virus. But some lethality can be tolerated, as far as the virus is concerned, as long as it is controlled—as long as Host A dies *after* infecting Host B. Then the virus can survive no matter what happens to the first host; the virus has taken up residence someplace else.

Consider the particular pattern of lethality of the rabies virus, which kills its mammalian host by destroying its brain. There is a beautiful logic to this virus's pathogenesis, or disease-causing process. The rabies virus, perhaps because of its highly efficient mode of transmission, has one of the broadest species ranges of any virus on earth; the exact same virus is capable of infecting bats, cats, dogs, squirrels, foxes, wolves, skunks, and human beings. It just so happens that the most efficient way for the rabies virus to be transmitted also inevitably kills most of the animals doing the transmission. (Bats, an important repository for the rabies virus, generally are not killed even when infected.)

"The virus goes into that part of the animal's brain that tells it, 'Thou shalt bite,'" says Bernard Fields, chairman of microbiology at Harvard. "The virus also goes into the saliva, and when the animal bites it is injected along with the saliva into the next animal's muscle." In the second animal, the rabies virus gets into the nerves by breaking into the neuromuscular junction, at which muscle and nerve intersect. And the cycle repeats: movement of virus into the biting region of the brain, movement also into the salivary glands, compulsion to bite, and passage through the bite into host number three. "The virus has to be lethal," says Fields, "or at least get into the brain, in order to get from Host A to Host B. If it doesn't have the bite, it [the virus] dies."

The way a virus causes disease depends on its tropism, or movement toward a particular host cell. Generally the connection is straightforward. The hepatitis A virus is tropic (pronounced *troe*-pik) for the liver; when it infects and kills enough liver cells, the symptoms of dark urine, malaise, nausea, and jaundice result. The rhinovirus is tropic for the lining of the nose and throat; it kills the cells there, and you have the miserable symptoms of the common cold.

Sometimes, it's not the virus at all that makes you sick; it's your body's attempt to fight the virus. Cell-mediated immunity carries with it some pretty damaging side effects: fever, pain, swelling, proliferation of toxins in the bloodstream. Animal experiments have shown that in many viral infections, if the animal's immune response can be aborted, no harm is done. When a certain type of mouse virus infects experimental mice, for instance, it usually invades the nervous system and leads to severe meningitis and rapid death. But scientists have saved mice infected with this virus—called LCM, short for lymphocytic choriomeningitis—by giving them drugs that totally suppress the immune system: no T cells, no inflammation, no response at all. If the immune system is suppressed, LCM causes no disease.

It's a natural question, when reading about something that seems so unrelentingly bad, to wonder how the first virus could possibly have come about. Evolution is supposed to be about survival of the fittest, and we like to think the direction of evolution is toward new and improved models. But here we are, highly evolved human beings, still being plagued by tiny primitive pathogens that—at least to our late-twentieth-century way of thinking—are no doubt far worse than their ancient forebears could possibly have been. Why, we might wonder, are there viruses—other than to bedevil all other living things, from flora to fauna, from microbe to man?

"Maybe they're just here, like the rest of us," says David Baltimore, co-discoverer of reverse transcriptase. "There's no reason anything has to have a function." But Baltimore is unusual in his apparent lack of curiosity. Most virologists, it seems, enjoy toying with ideas about how their object of study originated, and about its role in the universe. Ask a biologist what the true purpose of viruses might be, and a flood of theories is unleashed reminiscent of late-night bull sessions in college dormitories. It's one of those things scientists do best: musing about uses for apparently useless things, taking wild guesses about the origin of everything, pondering unanswerable questions while hanging around the laboratory during the inevitable downtime in scientific experiments.

One theory that has gained much credibility recently has been offered by Richard Dawkins, an ethologist (animal behaviorist) at Oxford and the author of several popular books on evolution. He proposes that viruses are "rebel human DNA." These prior pieces of ourselves "now travel from body to body directly through the air," he writes, "rather than via the more conventional vehicles—sperms and eggs.

If this is true, we might just as well regard ourselves as colonies of viruses!"

How did these rebel human genes break away? Dawkins leaves the details to those who are better versed in molecular biology. James Strauss, a molecular biologist at Caltech, has offered a few suggestions. Maybe RNA viruses were originally messenger RNA, he says, that overshot their mark while moving from the chromosome out into the cell cytoplasm during gene expression. And maybe DNA viruses originated as transposons, the "jumping genes" sections found on the human chromosome that tend to pull apart from the chromosome, move to a new position, and reinsert themselves elsewhere. As for retroviruses, Strauss believes they might have originated from the retroviral version of transposons, known as retro-transposons. These are the "jumping genes" of the DNA that codes for reverse transcriptase, and their DNA looks remarkably like the DNA phase of retroviruses.

It is possible, in thinking about how viruses came about, to wonder why anything *but* viruses exists. Dawkins chooses to view evolution as the survival of the fittest gene, not the fittest individual. While most biologists believe DNA to be an organism's way of perpetuating itself, Dawkins sees the truth to be exactly the other way around: an organism is DNA's way of perpetuating *itself*. So why did these "selfish genes" ever bother to collect themselves into organisms? His only explanation is that somehow a higher level of organization was better for the propagation of the gene.

Some scientists, when speculating about the origin and purpose of viruses, seem to go off into the realm of the phantasmagorical. One such theory, which it must be said is widely regarded as untrue, nonetheless makes for some fascinating speculation. British astronomer Fred Hoyle first achieved international acclaim—including a knighthood in 1972—for his then-radical theory about the origin of interstellar dust, which has since gained general acceptance. But in the late 1970s, he began venturing out into the field of biology, and announced with conviction that living organisms, and most prominently viruses, originally fell from outer space, and are continuing to fall from space even today.

As evidence, Hoyle pointed to epidemics that had erupted all at once at different spots around the world, such as an influenza outbreak in 1948 that even affected isolated Sardinian shepherds who never came in contact with other people. "I was trained to think . . . that a single contradiction is sufficient to upset any theory," he wrote. "This one single experience in Sardinia is sufficient to disprove the standard theory of influenza transmission by person-to-person contact, because solitary shepherds living for a long time alone could not have contracted the disease, all in the same moment, from someone else. To explain the facts, the influenza virus had to fall on the island of Sardinia from the air."

Unusual, yes. Idiosyncratic, yes. But there is one aspect of Hoyle's unorthodox theory that intersects with the ideas of many mainstream biologists. And that is that the existence of viruses, and their capacity to pick up genes from one host and pass them on to another, has played a central role in evolution. When viruses fall through the atmosphere and land on creatures great and small, writes Hoyle, they pass on genes indiscriminately. Some plants or animals incorporate those genes, others do not, but the presence of the virus seedings means that many different organisms can develop identical traits. Why else, he asks, would a flower and the butterfly that feeds on it have exactly the same yellow color? It is too farfetched to imagine that the slow and random process of evolution, as it is usually understood, would ever have allowed for such a perfect match. In Hoyle's mind, a virus carrying that yellow gene fell on both flower and butterfly, and the new adaptive feature of yellowness allowed each one to survive.

Leave out the part about outer space, and you have in essence one of the most intriguing theories about the purpose of viruses now percolating in the scientific community: the notion, put forth most eloquently by physician-essayist Lewis Thomas, that genes exist to stir up the genetic arrangements of species that otherwise would be staid, inflexible, and unable to adapt to environmental challenges. This is the "dancing matrix" image that has become the centerpiece of our investigation, conveying Thomas's idea that viruses "dart, rather like bees, from organism to organism, from plant to insect to mammal to me and back again . . . passing around heredity as though at a great party." This continuous flitting about, he speculates, might provide nature's most efficient method of keeping the gene pool varied and fluid, a way to keep "new, mutant kinds of DNA in the widest circulation among us." In his typically poetic way, Thomas has captured the prevailing theory about how viruses help propel evolution. Doing their cross-species jitterbug, viruses probably contribute to the on-going mixing of genes within a species, allowing that species the genetic flexibility required to take advantage of opportunities for adaptation. By increasing the rate of genetic recombination in the bigger creatures of the world, viruses are in effect helping those bigger creatures behave—evolutionarily speaking—more like viruses.

Survival of the fittest is the driving force of evolution, and it cannot occur unless some individuals in a species are genetically more fit than others. This happens through random events that give rise to variation in a species' genome: accidental damage to a gene, genetic mutation caused by environmental assault, the rearrangement of genes that have jumped from one region of the chromosome to another. If the change is debilitating, as many are, the offspring may die off rapidly. But if it conveys some survival advantage—a mutation, for instance, that enables an organism to get by with less water, in an environment undergoing a gradual decrease in average annual rainfall—then the offspring may thrive, the species may change, and evolution may take another small step forward.

Molecular biologists have recently been fascinated with just how the genetic rearrangement might occur. Some say viruses have had a hand in it. They cite as evidence the fact that, in piecing together the map of the human genome, they occasionally find familiar, identifiable gene sequences that look just like the genomes of certain viruses. The full complement of nucleotides in any organism is known as its genome, which can range in size from several thousand for an average virus to three billion or so for humans. These common sequences can be found in every human cell studied, and their presence seems unrelated to whether the individual shows any clinical signs of virus infection. Indeed, these viral genomes seem to be something people are just born with.

Malcolm Martin, chief of molecular microbiology at the National Institute of Allergy and Infectious Diseases, has an idea about why these viral genes appear from time to time within the normal human genome: it might be the human chromosome's signal to rearrange. Say you write out an alphabet in which each letter represents a gene. All twenty-six letters together represent the human genome. And say you deliberately insert a sequence of five *A*'s in a row to signal the beginning and the end of the genome. As Martin explains it, the genome would look like this: five *A*'s, twenty-six other letters, five *A*'s again, as in "*AAAAABCDEF . . .*" all the way to ". . . *WXYZAAAAA.*"

"The *A*'s are supposed to represent the retroviral genome," he says. "Genes can cross over every time they see a run of *A*'s." In this way, the viral genome is a sort of chromosomal square dance caller, prompting the occasional genetic do-si-do. Each viral sequence is a spot where the chromosome might break apart, letting the dancing genes know it's time to swing their partners.

The viral genomes scattered along the human genome all come from retroviruses. Because of the backhanded way in which they replicate, retroviruses are uniquely capable of integrating into the host's germ cells—the cells that develop into sperm and eggs—for direct transmission from one host generation to the next. In this way, so-called endogenous (inborn) retroviruses arise, essentially looking and acting just like normal human genes. Most endogenous retroviruses are quite innocuous, and quite ancient. Some of the retrovirus DNAs found in humans are identical to bits found in contemporary chimpanzees—and no doubt in the chimpanzees of eons ago. "The retroviruses we see in our genome today probably got into the germ line tens of millions of years ago as a result of exogenous infection," Martin says, "maybe in an epidemic similar to the one we're experiencing now with AIDS."

Endogenous retroviruses may serve several functions that actually help the host organism. One might be to make the host resistant to other viral infections. "Consider, for example, what happens when domestic cats are exposed to the baboon endogenous virus," writes Russell Doolittle of the University of California at San Francisco. "If the cat is carrying the closely related feline leukemia virus in its germline as a consequence of previous ancestral exposure, the cat is not afflicted. Cats not carrying the virus in their germline are smitten."

Another function of endogenous retroviruses might be in aiding in the routine operations of the cell itself. Before he started working full-time on AIDS research, Martin discovered a particular viral genome in the human placenta that might play a role in fusing cells between the lining of the uterus and the placenta. Millions of years ago, the virus might have infected a human and caused damage to the developing fetus. Today, the partial reawakening of that vestigial virus produces a protein that might be crucial for the healthy functioning of the organism. If this is true, then retroviruses have done more than propel evolution through the enrichment of the genome. They may actually contribute to the functioning of the organism itself.

The notion of viruses as the agent of plague and scourge is engrained in the popular imagination, especially in this age of AIDS, when one tiny virus can set a life tumbling into tragedy. The word itself, with a meaning that goes beyond simply a biological antagonist, enlivens our language. *Webster's New World Dictionary* defines *virus* as "anything that corrupts or poisons the mind or character; evil or harmful influence." In its most modern version, the virus has entered our vocabulary in a new way: a *computer virus*. A computer virus is a persistent glitch in the software, a booby trap that was somehow inserted deliberately by a mischievous hacker. Just like a biological virus, a computer virus is contagious; any computer that communicates with the infected program is infected as well. And just like a biological virus, a computer virus insinuates itself directly into the core of the program, becoming—until it is discovered and routed—a permanent part of the package.

This newest definition may be the one most youngsters will refer to as they encounter references to viruses—as a mechanical flaw in a man-made program. But at the same time that computer viruses are becoming popularized, our understanding of biological viruses is expanding to place these bizarre organisms in a new, central position in the biosphere—a position that may put viruses at the very core of life itself.

# PART TWO

# NEW THREATS

# 4

# *Mad Cows, Dead Dolphins, and Human Risk*

The week before Thanksgiving 1989, officials of the Centers for Disease Control in Atlanta thought that The Big One had finally arrived. A group of research monkeys in Virginia, which had been shipped just weeks before from the Philippines, was dying. The cause of death seemed to be the Ebola virus, one of the most deadly viruses known to man. If the monkeys were dying, the next likely step was that people would start dying too.

By November 13, five cynomolgus monkeys at the Hazleton Research Products facility in Reston, Virginia, were dead. By the twenty-seventh, the infection had killed seven more monkeys, which had been housed in a different room at the Hazleton plant. To stem the outbreak, and to protect the humans who might come in contact with them, all five hundred cynomolgus monkeys at the facility were killed.

As viruses go, Ebola is relatively new and incredibly deadly. It was first identified after outbreaks in the Sudan and Zaire in 1976. In the Sudan, where the disease struck between July and November, there were 150 deaths out of 284 infections, a death rate of just over half, making the virus about as lethal as yellow fever. And in Zaire, Ebola was even more devastating: between September and November, 318 people were infected, and a spectacular 88 percent died. A second, smaller outbreak, in the Sudan in 1979, had a death rate of more than 66 percent (twenty-two deaths among thirty-three infections). So when a Hazleton worker who had handled the sick monkeys developed symptoms of a viral infection on December 4—exactly one week after all the monkeys at Hazleton were killed—he was rushed to the isolation ward at Fairfax Hospital.

It turned out he had an ordinary case of the flu.

The dreaded Ebola epidemic in the United States never came to pass; maybe officials had caught it in time, or maybe the Asian virus infecting the monkeys would never have proved as virulent as the virus that had swept through Africa. In the end, a total of seven shipments of monkeys from three Philippine suppliers were actively infected; yet not one of the half a dozen or so people who had actually handled the monkeys, nor the nearly 150 who had been exposed to them en route from Asia, ever got sick. One worker, who inspected animals on their entry into the United States through New York's John F. Kennedy International Airport, was told over Christmas weekend that her blood sample revealed antibodies to the Ebola virus. But she seemed perfectly healthy, and she told the CDC investigators that her antibodies could probably be traced to an exposure to Ebola virus two years before.

The frenzy over the Ebola monkeys, in hindsight, might look like an overreaction. But it brings to light mankind's incredible susceptibility to the vagaries of viral traffic. The imminent prospect of a lethal virus so easily hitching a ride into the United States, exposing millions of Americans to an agent their immune systems had never before seen, made it clear—as if the AIDS epidemic hadn't made it clear enough— how fragile is the balance between microbe and man.

The monkey episode also raised some troubling questions, such as this: How did the Philippine monkeys get sick in the first place? Cynomolgus monkeys living in the Philippines don't normally harbor Ebola virus or any related member of the virus genus known as filovirus. Filoviruses are much more common among African monkeys, especially the rhesus and green monkeys. To determine where this surprising Philippine infection came from, epidemiologists at the CDC spent months tracking every step of the monkeys' trip from Manila to Reston. They thought at first the Philippine monkeys must have caught the virus from African monkeys that might have been on the plane alongside them. But this hunch proved wrong. "They were not next to any monkeys on the plane, infected or otherwise," says Susan Fisher-Hoch of the CDC. "The monkeys did not catch the virus in transit. They were infected in the Philippines." The big, unanswered questions are how this strange new virus behaves in the Philippines and how it changed—and when, and why—from its more deadly African ancestor.

The monkey story is a prime example of the cross-species transfer of a deadly virus, in this case from an African monkey to an Asian monkey—and, potentially, from an Asian monkey to an American scientist. Such trafficking across species boundaries, which occurs because two related species are in unusually close proximity, is an important source of emerging viruses. In general, this proximity can result from one of three conditions.

- Forces of nature, as when a drought or an earthquake leads animals to seek shelter in a new region.
- Forces of man, as when animals migrate because their natural habitat has been destroyed by forest clearing, air pollution, dam building, or the like.
- Forces of science, as happened with the Ebola monkeys, bringing into the laboratory setting already infected animals or their cultured cells.

Though the species hopping usually occurs from one animal to another, it occasionally also involves human beings. As shown earlier, one prominent theory says that AIDS originated as a disease endemic to African monkeys, which somehow made a jump across species boundaries and began infecting people. If the immunodeficiency virus did indeed cross species lines, it was simply following the traffic pattern of many other viruses in the past, and no doubt of many other viruses in the future as well. The goal here is to invoke common themes of viral traffic patterns, to provide a kind of bass accompaniment in the symphony of interspecies viral trafficking. Then once we get a better sense of the melody—that is, of the specifics involved in a particular virus leaping across two particular species—it should be an easy matter to insert the right notes and, ideally, to know which notes are coming next.

To generalize about how and when cross-species viral traffic occurs, scientists need lots of particulars. That's what Stephen Morse is doing these days: rounding up the particulars. Just as he once threw himself into collecting details about snakes and reptiles when he was a kid, Morse now has a new hobby: collecting virus stories. In his tiny office at The Rockefeller University, with just enough room for his bookshelves, computer equipment, and one perfectly organized file cabinet—and a sliver of a view of construction alongside the East River—he searches through obscure veterinary and biological journals for references to mysterious animal diseases. Morse still devotes most of his working day to laboratory work on herpesviruses—that is, after all, how he gets the federal grants needed to establish his scientific reputation, and to keep his job—but he spends more and more time in this little office, writing letters, composing speeches, editing books. In short, much of his spare time is spent in what he calls, with a touch of self-deprecation, "doing science with a word processor instead of a microscope." Morse's drive is propelled by a phrase he often quotes from Francis Crick: "One good example is worth a ton of theoretical arguments." What Steve Morse does these days is hunt down the good example.

"Most of the viruses we notice as emerging are not really human viruses at all," he says. "Their relationship with us may be accidental; we are hosts only by virtue of our proximity to the natural host." A lot of the viruses that cause humans the most trouble "can be found in natural hosts, including vertebrate hosts, where they cause apparently very little harm." They cause disease only when they infect an animal species for which they are not adapted. High lethality, in this view, is simply maladaptive for the virus, which depends entirely on its host's survival to ensure its own.

This change in host species no doubt accounts for the pathogenicity of many viruses. The morbillivirus that wipes out seals, porpoises, and dolphins probably resides in some other marine mammal with no ill effect. The parvovirus that now threatens dogs in every nation on earth might have started as a feline virus that was not nearly as devastating to cats. The callithricid hepatitis virus that stopped just short of traveling to an unexposed monkey population in Brazil has existed in Africa for years in chronically infected, but apparently healthy, monkeys. And so it goes, for the simian herpesvirus B that persists chronically in monkeys but can destroy a human brain in a matter of weeks; for the mad cow virus that seems to leap, with varying degrees of lethality, from sheep to cattle to antelope to cats; for the myxoma virus that was harmless to rabbits in Brazil and, at least for a while, quite devastating to rabbits in Australia.

Here we will investigate examples of cross-species viral trafficking not solely because they are stories with interesting twists and turns (though that they most certainly are), but because describing them and understanding the particulars are the first steps in developing a blueprint that explains how *any* virus crosses *any* species lines. If we study the zoonotic (animal) cases thoroughly enough, we may be able to find common themes that will help us understand how such viruses can change, sometimes with astonishing speed, into human threats.

In the fall of 1990, hundreds of dead and dying dolphins started to wash ashore along the Mediterranean. "They look normal, beautiful," said Jean-Michel Bompar, a physician from Toulon, on the southern coast of France. "But they shiver like a person with a bad flu, and they die quickly."

The flu analogy was apt, because these dolphins—much like a person with a bad flu—were suffering from a virus infection. The virus was new to dolphins, and was probably spreading because of the pollution in the Mediterranean Sea, which weakened their immune systems. A total of 700 dolphins washed ashore in Spain in 1990, and by the following summer, the dolphin virus had spread eastward to the coast of southern Italy. Italian scientists said the number of dolphins found dead—150 in Italy during July and August of 1991—only hinted at the true nature of the die-off. For every dolphin washed ashore and counted, they believed, one hundred dolphins might have died untallied farther out at sea. And the great fear among environmentalists was that the infection could spread farther eastward, toward Greece, where lived the last colony in the Mediterranean of monk seals, an endangered species whose numbers hovered at around two hundred.

The dolphin deaths were eerily familiar. Two years before the plague began, marine biologists witnessed a similar epizootic (animal epidemic) among harbor seals in the North Sea. More than eighteen thousand seals died in 1988, washing up for one terrible spring and summer along the shores of East Anglia, Ireland, Scotland, the Netherlands, and Germany. In April of that year, when the outbreak began, scientists from the Animal Virus Research Institute in Pirbright, Surrey, began a full-scale investigation of the death of the first few seals, whose tissues contained a virus that resembled foot-and-mouth disease virus, a picornavirus. (The name is a combination of *pico-*, a prefix that means tiny, plus *RNA*, to refer to the genetic material of the virus.) That was a frightening time. Foot-and-mouth disease is so deadly to cattle, and its virus so contagious, that it can be studied only in high-containment laboratories in isolated locations. Pirbright is one such location, and the Pirbright virologists were called in to action.

Foot-and-mouth disease turned out to be a red herring; the virus was detected in the tissue of the first few seals examined, but these rather crude techniques also isolated a herpesvirus, and it was not clear whether either one was even linked to the disease. By this time, though, some of the best virologists in Great Britain were on the case, and even without the foot-and-mouth disease threat there was no turning back. Brian Mahy, a virologist from Pirbright, and his colleagues collected blood from the dead and dying seals and used an assay called ELISA—an acronym for enzyme-linked immunosorbent assay—to check for antibodies to any known viruses. The only positive responses were to the type of virus known as the morbillivirus.

Different morbilliviruses tend to infect different species. The canine distemper virus infects dogs. The rinderpest virus infects cattle. The peste des petits ruminants virus infects goats and sheep. The measles virus infects humans. And now, it seemed, a previously unidentified morbillivirus was infecting seals.

The human measles virus is one of the most contagious viruses on earth. If you start with a few thousand susceptibles living together in a community—a "susceptible," in epidemiologists' lingo, is someone who has not acquired immunity through vaccination or through having recovered from an earlier bout of the same illness—then measles can tear through the population at a dizzying rate. If a single case of measles is introduced into this immunologically virgin territory, as happened in Greenland in 1951, within six weeks virtually 100 percent of the people will have caught it.

The Pirbright team, working with the Sea Mammal Research Unit in Cambridge, used most of the standard tools of an epidemiological investigation to determine the cause of the seal deaths. They arrived in East Anglia with blood-collecting equipment and took samples back to the high-containment laboratory in Pirbright. There they had tissue cultures for growing the blood samples, an electron microscope for taking a look at what they had, and a battery of antibodies with which to perform antibody neutralization tests.

ELISA tests and a similar assay, virus neutralization tests, are the heart of a study of this sort. If scientists don't really know what virus they are looking for, these tests can help narrow the range of candidates. Each depends on the ability of antibodies to bind tightly, or cross-react, with an antigen. In the ELISA test, if the type of virus in a serum sample is unknown, the virologist can mix it with a known antibody to which is attached a particular enzyme. This enzyme can then be activated, so that it begins a chemical reaction that yields a predictable color, and the depth of the color is compared against a standard chart. A deeper color means a larger quantity of antibody was able to bind with the antigens of the unknown virus—which is a clue as to which family of viruses the sample belongs, and just how close the relationship is likely to be. The greater the cross-reactivity, the greater the number of sites on the virus's outer coat that are well suited to the antibody—and, therefore, the more closely related the new virus is to the virus whose antibody is being used. Neutralization tests, similarly, mix a known antibody with an unknown antigen, but they measure the frequency of the binding with a different indicator: the presence on a culture dish of clear spots on an otherwise cloudy field. These clear spots, called plaques, are markers of an antigen-antibody complex. The number of plaques can be counted, just as the depth of hue in the ELISA test can be measured, to indicate how closely related the unknown virus is to a virus whose antibody is known.

This is, with only slight modifications, the way viruses have been discovered since the golden age of virology in the 1950s. At that time, The Rockefeller Foundation of New York was supporting six field laboratories on three continents, and scientists were finding "new" viruses everywhere they turned. In a single decade, the list of known arthropod-borne viruses grew from 34 (in 1951) to 204 (in 1959). The reason, of course, is that scientists suddenly had the tools for recognizing and categorizing viruses that had existed in nature all along. There were no new epidemics, no new pathogens; there were only scores of virologists out in the field or forest, catching animals and bringing them back to the laboratory for analysis.

When the harbor seals' blood was subjected to the ELISA test for morbillivirus, about half of them showed a cross-reaction with two of the strains: canine distemper and rinderpest. "This suggested that the seal virus shared [certain coat proteins] with both canine distemper and rinderpest," says Mahy. The other half showed a cross-reaction with another morbillivirus, peste des petits ruminants. These results helped narrow the field of possibilities—the seal virus was at least related to these viruses—but it made it clear that the virus they were looking for was not precisely any one of them. If the seals were dying of a previously known virus, they would have all reacted to the antigens present in one or another sample. As it was, the cross-reactions were strong but not 100 percent. This suggested that the seal virus was a type of morbillivirus but a previously unknown type. The next goal was to identify this new virus. "We are quite sure now, based on genetic sequencing, that we have found a fifth type of morbillivirus," Mahy says. "We call it seal plague virus, or phocid [seal] distemper virus." Mahy, who now works at the Centers for Disease Control in Atlanta, believes this is the same virus that infected the Mediterranean dolphins in 1990.

No one knows exactly why the new morbillivirus emerged when it did. It probably existed for years in some sea mammals with which, for some unspecified reason, the harbor seals had their first contact in the late 1980s. Significantly, when Mahy's team examined another seal species, known as gray seals, they found high levels of morbillivirus antibodies in their bloodstream. But the gray seals never got sick—a good indication that they may be the natural hosts of phocid distemper virus, serving as reservoirs without suffering any ill effect. Another theory is that the virus originated in the harp seals of Greenland, which had migrated farther south in 1988 than ever before. They, too, had antibodies in their bloodstream without any evidence of symptoms of seal plague.

At around this time several other small outbreaks of morbillivirus infection also occurred in the northern parts of Europe and Asia. Seals in Lake Baikal in Siberia died mysteriously in 1987; porpoises off the coast of Scandinavia experienced a virus epidemic in 1988; and sled dogs in northwest Greenland came down with what was, for them, an unprecedented epizootic of canine distemper in the winter of 1987–88. "The sea mammal population might contain a number of poorly characterized viruses that are in many cases closely related to known mammalian viruses," says Mahy. "This population, therefore, may be a viral reservoir of considerable potential significance."

And what about the Mediterranean dolphins that began dying off two years later? Obviously, something had to change in this region of the world, too, to turn this common virus into a killer. One thing that changed was the water itself. In the previous few years water temperature has increased in both the Mediterranean and the North Sea. Warmer waters encourage the spread of morbilli and other viruses. In addition, many of the pollutants in these seas—in particular PCBs (polychlorinated biphenyls)—can damage the immune systems of the seals and dolphins living there. Like humans with AIDS, immuno-compromised marine mammals are prey to deadly infection with microbes that would be harmless if their immune defenses were in good working order.

More recently, another flulike illness among marine animals has become noteworthy not only for the creatures that are getting sick but also for what their loss means to the ecological balance of the region. And in this case, the new threat once again can be traced to viral traffic.

In the spring of 1991, the population of sea urchins in the Caribbean experienced a rapid die-off. No one knew precisely how many urchins died, but experts said the population of black-spined echinoids—the most common sea urchins—went from one that could easily be spotted with the naked eye to one that was all but invisible in the coastal waters off Puerto Rico and the Florida Keys. "We may be witnessing the largest extinction of a marine animal ever recorded," said Robert Bullis, director of marine animal health at the Marine Biological Laboratory in Woods Hole, Massachusetts. The symptoms of the urchin epidemic resemble those of other viral diseases. Urchins become lethargic; their spines droop and eventually fall off; they stop eating, stop moving; and they ultimately waste away. Bullis, who conducted autopsies on dozens of urchins sent by overnight mail to his Cape Cod laboratory from Florida, said the internal organs of infected sea urchins, which are usually brightly colored, turn to brown mush.

The loss of sea urchins in the Caribbean reverberates throughout the food chain and imperils the survival of the magnificent coral reefs that attract thousands of tourists every year. The sea urchins maintain the balance on the reefs by feeding on the algae that grow on the corals. If the sea urchins are not there to keep algae growth in check, the algae can grow so profusely that the corals are smothered.

How did this illness begin? Bullis believes it might have been through a virus carried in bilge water of a ship traveling from the Pacific Ocean to the Caribbean via the Panama Canal. Bilge water is taken into the ship's hull from the surrounding ocean, and carried as a kind of ballast as the ship makes its journey. By dumping bilge water into the Caribbean, a ship traveling from the Pacific introduces microbes that have never before been in that marine environment. And in this new environment, the theory goes, the Pacific Ocean virus found a highly susceptible, immunologically naive population to infect—much as the smallpox virus brought by the European explorers found a new population in the American Indians.


The Scottish veterinarian was brokenhearted to see the succession of dead dogs brought into her laboratory at the University of Glasgow. "Young pups, apparently fat, fit and healthy, were suddenly 'dropping dead' from heart failure," read an account of the report from the vet, Irene McCandlish, about the puppies that died in that first wave of the epizootic. It was October 1978. She performed autopsies to see if anything looked odd about these dogs, which had been frisking about just moments before they stopped, shuddered, and died. In all the heart tissue she examined, McCandlish found particles that resembled parvovirus, a tiny virus previously thought to infect only mink, raccoons, and cats.

Curiously, another dog disease also appeared in Scotland at the very same time. Older dogs were coming down with a highly virulent illness: a profuse, foul-smelling diarrhea; vomiting of a frothy fluid stained with bile; and rapid dehydration. The disease raged through entire communities, infecting a high proportion of dogs, killing about 10 percent of those infected. The dogs that died usually did so rapidly, within seventy-two hours of their first symptoms. More curious still, an identical pairing of these two previously unknown canine conditions—heart failure (myocarditis) in puppies, severe diarrhea (enteritis) in older dogs—was arising simultaneously on at least three separate continents. In the United States and Australia, the disorders were first reported in August 1978; in Canada, Europe, and Great Britain, reports started coming in during October. (Recent retrospective reviews of frozen sera from dogs that died of mysterious illnesses—the kind of reviews that helped scientists confirm very early cases of AIDS—indicate that the earliest case of canine parvovirus probably occurred in Greece in 1974.) In all of these outbreaks, the same parvovirus was recovered from the tissue of infected animals.

Parvoviruses are among the smallest and simplest viruses known, made up of a short length of DNA with just enough genes to direct the production of four or five proteins. Three of those proteins are needed to form the virus coat, so there's not much left over for other life functions. The parvovirus must corral materials already in the cell to supply everything it needs to make more virus—or, to be more precise, everything it needs to get the cell to make more virus. Unlike many larger viruses, which often package or produce at least a few of their own enzymes, the parvovirus can reproduce only in cells that are actively dividing. Nondividing cells do not contain enough of the enzymes needed for viral replication, transcription, and gene expression.

This penchant for dividing cells explains the observations that Irene McCandlish was puzzling over in 1978. Why, she wondered, did parvovirus infection cause myocarditis in young puppies but diarrhea in older dogs? What did the cardiovascular system and the gastrointestinal system have in common? McCandlish had a theory, which she presented to her colleagues at the annual British Veterinary Association meeting in September 1979. No matter when in a dog's life infection occurred, she reasoned, the parvovirus would replicate only in those tissues where the body cells were dividing rapidly. During a puppy's first week of life, many rapidly dividing cells were in the heart, which was still growing. So when the infection occurred in the first week after birth, the result was myocarditis. After a few weeks of age, when the heart had reached its mature size, the fastest-dividing cells were in the intestine, where the epithelial cells lining the intestine were constantly being replenished; at this point, a canine parvovirus infection led to enteritis. Between the ages of one and five weeks, which McCandlish thought of as a transition period, dogs might exhibit a little bit of both: primarily enteritis, with small amounts of myocarditis. She believed that in these cases, the parvovirus had reached the heart too late to find and invade enough dividing cells to do any real damage.

Parvovirus is ideal for laboratory study because it is so tiny; indeed, the name *parvo* comes from the Latin for "small." Since it is only about five thousand nucleotides long—compared with a larger virus, such as poxvirus, which can be as much as thirty times longer—parvovirus is relatively easy to sequence. *To sequence* is a verb that means, to a molecular biologist, just what the noun implies: to map out, in consecutive order, exactly which nucleotide comes first, second, third, or five-thousandth in the genome. Because of how thoroughly virologists now understand the little parvovirus, they are able to manipulate and reconfigure it in ways that would be impossible with more complex viruses. This means they can, if they want to, actually imitate the steps in nature that might have occurred during the change from the cat parvovirus into one that infects dogs.

At Cornell University, a team of veterinarians and molecular biologists managed to do just that. Colin Parrish and his colleagues began by creating DNA clones, or colonies, of both types of parvovirus: the cat version, called feline panleukopenia virus, and the dog version, called canine parvovirus 2. They then combined the two genomes into a variety of hybrids, like the children's card game in which different animal heads can be switched around to sit atop different animal bodies. If they began with the 5,000-nucleotide-long cat virus, they replaced portions with the same region from the dog virus to see what changes would make the cat virus behave more like the dog. It was like starting with a line of 5,000 interlocking toy "pop beads" that were all red, and then sticking in strings of yellow beads from a second line of pop beads that were all yellow. So you'd end up with a red chain except for, say, 400 yellow pop beads in position numbers 305 through 704, or a red chain except for 84 yellow beads from position numbers 3,229 through 3,312. Then you could see whether any of your red-and-yellow hybrids had the properties you were studying. By finding which hybrids had which properties, you could make inferences about where the yellow-bead insertion made a difference—and from there, you could make a determination of which region of the genome controlled a particular viral function.

In this way, Parrish and his colleagues created a variety of recombinants. They could compare each one to canine parvovirus, which they had already isolated from sick dogs and sequenced, nucleotide by nucleotide. With one of these recombinations, they found what they were looking for. A small portion from the canine parvovirus genome added to the original feline sequence—in other words, just a little bit of yellow among the red—made the new virus, in culture, behave like the viruses that infected dogs in nature. The critical region was only 730 nucleotides long—less than 15 percent of the entire feline virus genome. And yet, in vitro and in vivo (that is, in the laboratory and in a live animal), that hybrid cat/dog virus behaved almost exactly like canine parvovirus. It bound to antibodies specific for canine parvovirus. It grew well in cultures of dog cells. It even grew well in dogs themselves—something that the pure feline virus was unable to do.

What Parrish had done on his experimental pop bead chain, it seemed, was to turn a chain that began as a feline panleukopenia virus into one that functionally resembled canine parvovirus. Not only was this useful in making some predictions about the particulars of how canine parvovirus emerged, but it also offered clues to the general plan of host specificity of parvoviruses—and perhaps of other viruses as well. Parrish's work showed, as he points out, "that most of the host range properties of the virus are encoded by a short region within the capsid [coat] protein gene."

Parrish's laboratory earned a national reputation for its work with canine parvovirus and developed a collection of antibodies against the antigens, or proteins, on the coats of both the canine and feline parvoviruses. Veterinarians from across the United States began to send feces samples to Ithaca for confirmation that it was indeed parvovirus that was infecting local dogs. The antibodies were of a particular type called monoclonal antibodies, meaning they were all derived (or cloned) from a single (mono-) antibody. If a sample sent in from a local vet did in fact contain canine parvovirus, it would react with the monoclonal antibody in a highly predictable way. "As we began to characterize these field isolates using our monoclonal antibodies, something surprising turned up," Parrish recalls. "The viruses isolated from samples collected after 1980 all appeared to be antigenically different from the viruses that were collected before that year." He made monoclonal antibodies from those later virus samples and found that the viruses differed slightly, in terms of nucleotide sequence and coat antigens, from the original canine parvovirus. This suggested that, somewhere around the beginning of the decade, a new strain of canine parvovirus had arisen.

Parrish solicited the help of collaborators from around the world—from Australia, Belgium, Denmark, France, Japan, and the United States—to conduct a historical review. These scientists all used Parrish's series of monoclonal antibodies on virus samples they had preserved in their own laboratories, in an effort to see when the second type, which Parrish was calling canine parvovirus 2a, had emerged. It turned out that it had never been seen before 1979, but that by the mid-1980s canine parvovirus 2a had largely displaced the original canine parvovirus 2 around the world. More surprising still, a *third* type of parvovirus, which Parrish calls canine parvovirus 2b, started to emerge around 1984. By 1990, canine parvovirus 2b had become far more common, at least in the United States, than either canine parvovirus type 2 or type 2a.

"What were the selective forces that gave CPV-2a such an advantage?" Parrish wonders. And why did 2b displace 2a? The answer is still obscure—as is the answer to the question of where canine parvovirus 2 came from in the first place. In the early 1980s, one popular explanation was that canine parvovirus originated as a mutant derived from a live feline virus vaccine. The cat virus, according to this theory, made a small error in its replication in culture, in a region responsible for host specificity. The mutant would have died out altogether if it was forced to pass from one cat to another under natural conditions, but in culture dishes filled with cat tissue cells, being passed from one cell culture to another by a person with a glass pipette, the mutant virus survived. Those few genetic changes, which made the virus capable of infecting dogs, were included in the cat vaccine used around the world, and dogs were subsequently exposed to a new infectious agent.

Parrish was one of the original proponents of this theory, but he has since discounted it. He does not like having it repeated either, even in a historical way. "I think that it is an example of a good idea that was floated by people who subsequently never took the time to examine whether it was correct or not," he says. "After a while such ideas become transmuted into 'facts,' and then they become dogma, and then if it turns out that the idea is wrong, it becomes difficult or impossible to propagate the real truth in the face of the already existing, incorrect 'truth.'" One reason this particular "fact" has taken hold so firmly—besides its being rather simple to explain, and putting human beings at the center of the veterinary action—may well be that it appeals to scientists' and science journalists' sense of irony. The nice twist to the story is this: since the mid-1980s, dogs around the world have been given a vaccine to protect them against canine parvovirus, resulting in the virtual elimination of the myocarditis that killed puppies early in the epizootic. If it were true that the canine parvovirus originated because of a mutation that proliferated in a cat vaccine preparation, how fitting it would seem now to be protecting puppies with a vaccine that is, with only slight modifications, the very same vaccine that created the problem in the first place.

Science is not an O. Henry story, of course, and there are now several more probable explanations to account for why canine parvovirus emerged in the late 1970s. The dog virus might have begun with a direct transmission from cat to dog or by transmission from some other intermediate species whose identity is yet to be discovered. But these scenarios are routine and rather dull; they have none of the spark that enlivens the story about vaccine mutants. That, perhaps, is why— to Parrish's great dismay—that explanation is likely to be repeated until a better one comes along.

Even if vaccine manipulations have no proven role in passing viruses from one animal species to another, another human intervention probably does contribute to the cross-species transfer of pathogens: zoological parks. When animals from different regions of the world are housed in a zoo of a few acres, they often encounter conditions prime for passing viruses to and fro. Occasionally what that means is that animals from one hemisphere, which might have learned to coexist with indigenous viruses, infect their zoo neighbors—with ancestry in a different hemisphere—with viruses against which they have no immunity. The resulting epizootics, in which dozens of zoo animals might die, are bad enough. But with some zoos now engaged in a venture to reintroduce captive-bred endangered species into their natural habitats, these emerging viruses could well break out beyond the confines of the zoo itself and threaten native animals that are already struggling to survive.

The reintroduction effort coordinated at the National Zoo in Washington, D.C., often leads to some surprising sights. In the woods just off the well-traveled paths between exhibits, visitors might chance upon several golden lion tamarins—delicate monkeys, only about a foot long, that are named for their lush manes and their beautiful tawny coats—prancing along the metal railings, sneaking behind trees and into the glades, coming surprisingly close to the human legs hurrying by. Earnest young zoo volunteers follow closely behind, recording the monkeys' every move on their note pads, too preoccupied to say anything to passersby other than "Watch out!" This odd behavior is the first step in the zookeepers' effort to save the golden lion tamarin from extinction. The idea is ultimately to move the monkeys from the only home they have ever known—the confines of a zoo cage, where meals are delivered regularly and all predator species are on the other side of the bars—to the home of their ancestors.

The golden lion tamarin is "the most beautiful monkey in the world," according to journalist Diane Ackerman, "a sunset-and-corn-silk-colored-creature, about the size of a squirrel, that lives nowhere else on earth" but in the vanishing tropical rain forests of Brazil. The entire worldwide population of these endearing, human-faced creatures has shrunk to just four hundred, leaving zoos as a valuable repository of the species. That is why zoos around the world, along with scientists at the Rio de Janeiro Primate Center, are engaged in a collaborative experiment to reintroduce these endangered animals into their native habitat.

Zoo officials know the risks of reintroduction: the captive-bred monkeys might be unable to forage successfully; they might become disoriented in the tree canopy hundreds of feet above the forest floor; they may suffer the breakup of the family units they have always known. (Golden lion tamarins are quite loyal to their families, and they mate for life.) Since the Golden Lion Tamarin Conservation Program was begun in the mid-1980s, with reintroduction as its bedrock, seventy-five monkeys have been reintroduced into the Amazon rain forest. But only twenty-seven are still alive: about two dozen have succumbed to disease, and another two dozen, presumably, either starved to death or were killed by predators.

These are the risks of any conservation program, and officials involved have been aware of them from the beginning. But it was not until January 1991 that they came face to face with another risk they had not fully considered before. By bringing zoo-bred monkeys into the rain forest, in the hopes that they will mingle and interbreed with wild Brazilian tamarins, scientists may be endangering not only the zoo monkeys. They may be endangering, too, the tamarins that already live in the wild.

A virus of some kind has been rippling through the captive monkey population of the United States since the beginning of the 1980s. It was known as the callithricid hepatitis virus, or CHV: callithricid for the type of animal it infected (the callithricidae family of monkeys), hepatitis for its most characteristic symptom (liver inflammation). In 1990 alone, nearly a dozen American zoos experienced epizootics of

CHV among their New World monkeys; a total of sixty-five marmosets and tamarins died. One of the zoos experiencing an outbreak of monkey virus that year was the Brookfield Zoo, located about a half-hour's drive from downtown Chicago.

Soon after the outbreak, in the summer of 1990, the Brookfield Zoo sent six golden lion tamarins to the National Zoo to be prepared for reintroduction to Brazil over the winter. In the Brookfield shipment were an adult male, an adult female, and their four offspring. When they were first examined as part of the routine intake procedure in Washington, the male, whose name was Flash, showed signs of CHV antibodies. This was not surprising, since he had survived the recent Brookfield outbreak.

Flash was held for observation—a standard procedure for all golden lion tamarins en route to Brazil, whether they seem to be carrying disease or not. Tamarins from the six other American zoos engaged in reintroduction are always quarantined in Washington for several months, so they can be tested for signs of unusual infections or genetic problems; about 15 percent of the monkeys, for one reason or another, never make it to Brazil. Those that pass are flown to Brazil by National Zoo staff members, who make an annual pilgrimage to hand-deliver the animals. In Brazil, the tamarins are kept in a protected forest preserve and closely observed for six months before they are released to the wild, where they finally will come in contact with native golden lion tamarins.

During his quarantine, Flash never got sick. Neither did the rest of his family group. In fact, the others in his family never showed signs of the antibodies that Flash had. The zoo staff decided it would be safe to send Flash to the Amazon with the rest of the monkeys scheduled to depart in January. But just days before takeoff, they changed their minds—because the virus Flash had been exposed to had been more carefully characterized. The callithricid hepatitis virus, they discovered, was closely related to the mouse virus LCMV (short for lymphocytic choriomeningitis virus), one of a large group of rodent-borne viruses known as arenaviruses.

The relationship to LCMV was terribly worrisome. LCMV is an endemic mouse infection that is known to persist, usually with no harm to the mouse, for an entire lifetime. It is usually passed from mouse to mouse, probably by being shed in the mouse's urine or feces and then combining with dust that can easily be inhaled. But occasionally another species gets in the way of LCMV and develops a devastating illness. When a tamarin or marmoset contracts the virus, most likely by eating infected newborn mice that were (until this finding) a staple of many zoo diets, it runs about a 50 percent chance of dying. "Now that we knew what virus group it was in, the question was, Is this persistent in primates?" says Richard J. Montali, the pathologist at the National Zoo. "You can usually recover LCM virus from infected mice. But we didn't know—and we still don't know—if you could recover the LCMV-like virus from primates that survived the infection." If the callithricid hepatitis virus *did* remain in Flash's tissue—quite possible, given the family of viruses it came from—zoo officials worried about introducing it into the fragile ecosystem of the Brazilian rain forest. Imagine the possibilities if Flash were sent there with the callithricid hepatitis virus still in his body, and still potentially infectious. Say Flash was brought into the jungle and died. If his carcass were eaten by a rodent, and then *that* rodent were eaten by another rodent or a small mammal, the virus might begin a cycle in the rain forest, subjecting not only tamarins but possibly other susceptible animals to an emerging virus that originated in a captive environment. And the zoo, in the name of preservation of a species, might actually be engaged in a far greater iniquity: introducing a disease potential that could strike animals in the wild.

This, of course, is the great paradox of reintroduction campaigns. Zoo officials have always known that they were taking some risks with reintroduction—risks that still have not been fully examined. In this particular case, callithricid hepatitis virus is not literally a new virus; it is very closely related to LCMV, which probably exists—as the mice that harbor it do—on nearly every continent on earth. But its presence in the monkey named Flash did point out to zoo officials that they are running the risk of introducing novel organisms into countries where they have never been before. They believe it's a calculated risk, one worth taking for the greater good of maintaining the genetic diversity of the biosphere. And as Benjamin Beck, who directs the tamarin program at the National Zoo, is fond of pointing out, "I think the reintroduction of tamarins into the jungle will bring in far fewer pathogens than does the constant reintroduction into that same region of human beings." Zoo reintroduction programs are just another route, albeit an unusual one, for the increasing proliferation of viral traffic that we are now hearing about again and again.

Act I, Scene I. The barren Utah landscape is dominated by the headquarters of BioTek Agronomics, which looks for all the world like an ordinary agricultural research firm. This is the opening shot for *Warning Sign*, a 1985 movie starring Sam Waterston. Within the first ten minutes, the truth about the company is revealed: BioTek is secretly engaged in developing products for biological warfare. After a fluke laboratory accident, in which a scientist steps on a glass vial containing a genetically engineered virus, it seems no one in the whole state of Utah is safe. The fictitious accident exposes scores of lab workers to the mutant virus, which turns against these men and women in just the way it was designed. Already quite lethal even before it was manipulated, the genetically altered virus now has a nasty ability to convert to aerosol form and, like influenza, spread easily through the air, passing from one person's respiratory tract to another's with every breath. "It drives people crazy," explains a virologist as the movie reaches its bloody climax. "Soldiers murder their comrades, and then they die."

The bad guy in *Warning Sign* is the brilliant scientist directing the germ warfare research. He is one of the first to be exposed in the lab accident, and his own genetically engineered creation makes him go totally berserk. The character's name is Nathan Nealson—a play on the name of real-life virologist Neal Nathanson of the University of Pennsylvania. This anagram is no coincidence; it was a nod to Nathanson in appreciation of the small role he played in the creative genesis of the movie. Nathanson does not want to talk about his hand in *Warning Sign*—and judging by the dubious quality of the film, it's no surprise—but he will say that he did indeed give director Hal Barwood the idea of a particular virus that is "a really bad virus you wouldn't want to get out of the microbiology laboratory." Beyond that, he offers no comment. "It was a cockamamie idea," he says, "just idle cocktail party chatter."

*Warning Sign* never became a box office success, but its very creation shows how prevalent is the idea that laboratories are dangerous places where bizarre manipulations are going on. When the manipulations involve viruses, they seem even more pernicious. Strange viruses erupting from some mad scientist's lab bench tend to populate our collective nightmare.

But it's not just anti-scientific Luddites who question the wisdom of too much laboratory manipulation. Even respected virologists wonder whether by the very act of raising viruses in culture, scientists are creating new strains that will prove more dangerous than the strains from which they originated. That was the possibility invoked by one of the nation's leading AIDS researchers, Robert Gallo of the National Cancer Institute, when he wondered in early 1990 whether experimentation on the human immunodeficiency virus could eventually make the virus more pathogenic. The reason for his concern was the growing use of a new strain of experimental mouse: a mouse with no immune system of its own, into which human immune system cells are transplanted. In biologist's lingo, this model is known as the SCID-hu mouse: *SCID* for "severe combined immune deficient"; *hu* to indicate the mouse's receipt of grafts of human cells. The SCID-hu mouse is a tiny live model of a functioning human immune system. With HIV injected into these mice, Gallo wondered, could it come in unnatural proximity to some *mouse* retrovirus? Could that proximity result in a rearrangement of viral genes, something that inherently promiscuous retroviruses are prone to? He and his colleagues, led by Paolo Lusso at the National Cancer Institute, designed an experiment to see what would happen if both HIV and murine leukemia virus—an endogenous retrovirus of mice—co-infected the same human immune cells in a culture dish. And in cell culture (which is, of course, quite a different environment from a living SCID-hu mouse), the two viruses did indeed exchange bits of genetic material. The new HIV strain grew faster than other HIV strains, and was able to infect the cells lining the respiratory tract, which usually resist HIV infection. This last was especially troubling: infection of the respiratory tract is one of the requirements for a virus to be able to be transmitted through the air.

"Interaction between endogenous retroviruses and other viral agents can determine phenotypic alterations [changes in the surface of the virus]," wrote Lusso, Gallo, and their co-workers, "or even genotypic recombination, resulting in permanently modified viral progeny." Because of this tendency to mix, they wrote, any animal models for AIDS should be approached with caution—both in terms of the validity of whatever findings seem to result, and for safety reasons, "in consideration of the potential in vivo generation of more pathogenic phenotypic or genotypic variants of HIV-1."

Others were not worried. Stephen Goff, a molecular biologist at the College of Physicians and Surgeons at Columbia University, told a *New York Times* reporter that he doubted that picking up pieces of the mouse virus would make HIV more resistant to the things that usually kill it, like drying or detergents. And Mark Feinberg of the Whitehead Institute for Biomedical Research at the Massachusetts Institute of Technology said the mouse virus proteins that stick to the AIDS virus may allow HIV to enter a new group of cells, but that happens only once, because the proteins don't fundamentally change the AIDS virus. It would revert to its original form when it replicated.

Endogenous viruses create a laboratory hazard whenever animals are used for experiments—especially animals like monkeys and chimps, whose endogenous viruses are likely to be able to infect humans with little trouble. These events do not involve deliberate or inadvertent rearrangement of the viral genome; all that need happen is that a human researcher gets unnaturally close to an infected animal or even to the animal's cells. Among the most deadly of these accidental transmissions, which have killed a few dozen laboratory workers in the past fifty years, are two viruses that any biologist knows by reputation: the Marburg virus, which erupted in the German laboratories that were handling African green monkey cells to make polio vaccine; and simian herpesvirus B, also known by the shorthand herpes B.

The most recent herpes B outbreak in the United States involved four individuals, two of whom died, from the naval Aerospace Medical Research Laboratory in Pensacola, Florida. The first victim, a man of thirty-one, had been an animal handler for eight years when a monkey finally bit him. Although he often failed to wear the thick leather gloves that are required whenever a worker touches monkeys, investigators who were later called in from the Centers for Disease Control were not able to determine whether he was wearing gloves on March 4, 1987. But they do know this: it was on that date that a sick rhesus monkey—ill with diarrhea and severe conjunctivitis of both eyes—bit the young man on the left thumb. By March 9, the man's left arm was numb. By March 22, he was seriously ill: he had lethargy, fever, chills, dizziness, and muscle pain. After his numbness spread to the entire left side of his body and he started seeing double, he was admitted to the hospital on March 28. He was placed on a respirator, treated first with intravenous acyclovir (a potent antiviral drug) and later with an experimental antiviral known as DHPG. But he eventually went into a coma, and never regained consciousness.

The same day that the animal handler was admitted to the hospital, his colleague, a thirty-seven-year-old biological technician at Pensacola, was hospitalized with similar symptoms. In retrospect, investigators concluded the technician may have been bitten or scratched by the same monkey that bit the animal handler. But this is essentially conjecture. The only thing that federal epidemiologists know for sure is that the technician received some sort of penetrating wound on his left forearm on March 10, that in the previous weeks he had frequent contact with that sick rhesus monkey, and that during his thirteen years with the facility he sometimes failed to wear protective leather gloves. The man developed herpeslike lesions on his left arm, and on March 26 he saw a dermatologist who diagnosed it as herpes zoster and prescribed a topical cream of acyclovir. The man never filled the prescription. He just stuck with an over-the-counter hydrocortisone cream, which his wife had been applying to his wound for the previous week. Soon, though, the technician grew terribly sick. He developed numbness in his left arm, chest pain, fever, difficulty breathing and swallowing, confusion, lethargy, double vision. On March 28, shortly after he was hospitalized, he stopped breathing; he was placed on a respirator. Despite a month of intensive treatment—first with intravenous acyclovir and later with DHPG—he died.

While her husband was deteriorating in the hospital, the technician's wife also became sick. From applying hydrocortisone cream to her husband's forearm, her own fingers became infected; in one spot, underneath a ring, she had a rash that was so itchy that she scratched it until it bled. On April 1, a dermatologist prescribed oral acyclovir for her rash. One April 7, the laboratory report on a skin culture from that rash came back with startling news. The woman, like her husband and like the animal handler, was infected with the monkey virus known as herpes B.

Herpes B is the monkey equivalent of human herpes simplex, the kind that is generally harbored with little effect until it erupts now and then into a cold sore. Like most other herpesviruses, herpes B is highly adapted to its natural host. It usually causes no symptoms in monkeys, residing quietly in skin and nerve cells and replicating there—a situation that could be considered ideal, at least from the virus's perspective. When herpes B crosses species lines, though, it becomes highly virulent. In humans, it can wipe out a victim's brain in a matter of days. In the fifty years since herpes B was first identified—by none other than Albert Sabin, the developer of the polio vaccine—it has been arguably the most dreaded occupational disease among veterinarians, animal researchers, and laboratory technicians. Mercifully, it is quite difficult to catch; since the 1930s, only twenty-three cases of symptomatic human infection have been reported in medical journals. But the virus is highly lethal. Of those twenty-three cases, eighteen (nearly 80 percent) ended in death.

It is no surprise, then, that physicians bolted into action when the wife of the infected biological technician was diagnosed with herpes B. Hers was the first known case of human-to-human transmission of the virus; in all other instances, victims had been bitten or scratched directly by an infected monkey. The wife, who was twenty-nine years old, was hospitalized on April 7 and given intravenous acyclovir. By then, her eyes had also become infected, probably from handling her contact lenses. But the condition was caught in time. The young woman improved with the medication, her rash cleared up, her eye involvement never became a problem. Doctors took fluid samples from her mouth and eyes every three or four days to see whether herpes B was still there. By the time her husband died, on April 28, her fluids were free of infection.

There was one more person involved in the Pensacola outbreak, and his story is perhaps even more troubling than the story of the technician's widow. This man, a fifty-three-year-old laboratory supervisor, never broke guidelines; he always wore leather gloves over rubber surgical gloves when he was catching a monkey, and he kept the surgical gloves on the whole time he handled the animal. The rhesus monkey he was working with on March 11 was perfectly healthy, and the man reported no untoward encounters: no bites, no scratches, no contact at all with monkey body fluids. Yet on March 27, some spots on the middle finger of his right hand started to itch; by March 30, they were dry and crusted over. Knowing the fate of his colleagues, the laboratory supervisor acted quickly enough to catch the infection in time. His infection was confirmed as herpes B, and he began intravenous acyclovir on April 10. He was declared free of virus, except for a single rectal culture, by May 8. The supervisor's case was of concern because the mode of transmission was difficult to track. The monkey, first of all, had been healthy; there was no reason to be on alert for the possibility of viral transmission. In addition, nothing about the man's encounter with the monkey suggested that a blood-borne virus like herpes B would even have the *chance* to be transmitted—there were no bites or other ways for the man and the monkey to exchange blood. And everything about the way the supervisor had conducted himself—with the possible exception of failing to anesthetize the monkey before handling it, which is often recommended as a way to avoid bites and scratches—was according to the letter of the laboratory's own regulations.

The Pensacola outbreak ended the way most herpes B outbreaks have in the past: with a few workers dead but no serious secondary transmission. It seems that once herpes B gets out of the monkey host to which it has adapted, it dies out. Considering how devastating the infection proves to be to the few individuals who are directly exposed, this seems lucky indeed.

Other laboratory hazards from animal viruses continue to arise. Hantavirus, whose presence in laboratory rats is what first alerted Stephen Morse to the possibility of emerging viruses, have proved to be more dangerous in the laboratory than they even are in nature. According to James LeDuc, who worked with Hantavirus in his lab at the United States Army Medical Research Institute for Infectious Diseases in Fort Detrick, Maryland, the biggest risk of Hantavirus comes from the fact that it can be transmitted through the air. In clean, carefully contained laboratories housing large numbers of rodents—not even infected rodents necessarily, but sometimes just rodents from the wild—he says evidence exists of rather easy transmission of Hantavirus. "We know of cases in which people walked in and spent very short periods in the lab, doing nothing more than breathe the air, and they've become infected," LeDuc says. At Fort Detrick, efforts are made to avoid this inadvertent exposure by strict control of airflow in the Hantavirus laboratory. LeDuc's lab has a well-controlled ventilation system, a rack for rodent cages that sits inside a laminar flow case with unidirectional airflow (always flowing out), and a requirement that any time a rodent or a Hantavirus sample is handled, it is always worked with under a hood with its own air source separated—by an invisible wall of moving air—from the air in the room itself.

These precautions probably should be taken in any laboratory working with rodents, not just in Hantavirus laboratories—just as Joshua Lederberg suspected. "We know of many lab rats that have Hantaan virus," says LeDuc, "even when they are commercially supplied and even when the lab isn't studying Hantaan." The inadvertent infection of rats bred specifically for laboratory use is a continuing problem, he says, especially in Japan, South Korea, Europe, and probably China. Because urine and feces tend to accumulate in any laboratory, and because so many rodents are usually housed in one enclosed room, the infection of laboratory workers through airborne transmission is a problem as well.

Any laboratory worker who handles highly lethal viruses can run into trouble. Even if the viruses are not especially deadly, even if all containment guidelines are carefully observed, surprises might await —as happened to the laboratory supervisor in Pensacola. But a risk for a laboratory worker, disturbing as it is, is one thing; a risk for others in the community is something else again. Scientists and laboratory technicians have chosen careers they know carry a certain danger; for many, that is part of the appeal. "Accidents happen," says Robert Tesh, a scientist at the Yale Arbovirus Research Unit. "If you're a chemist and you splash yourself, usually you can wash it off and it's no big deal. But if you're an infectious disease expert and you splash yourself, you might get really sick; you might even die." Most virologists seem to know and understand those risks, and have chosen to accept them for the chance to conduct scientific work that is challenging, meaningful, and, yes, even a bit exciting.

People living in the community have made no such choices. And scientists and community leaders alike seem to agree that it is inappropriate to burden the public at large with even the possibility of a new virus escaping from the lab. Such an accident would confront communities with risks the magnitude of which cannot be fairly calculated.

When these eruptions have occurred, public action has worked speedily to restrict the chance of a repetition. In 1978, when a medical photographer contracted smallpox through a ventilation duct while taking pictures in a laboratory one floor above a smallpox laboratory in England, the international scientific community responded to the tragedy by prohibiting any research involving variola, the virus that causes human smallpox. But even without any ongoing research on variola, smallpox still is considered a risk—albeit a tiny risk—to individuals in a few occupations. Military personnel, for instance, would be in the line of fire if smallpox were to be used in biological warfare, a tempting possibility for some evildoer, since people in this post-vaccination period are currently as immunologically naive with regard to smallpox as were the American Indians in Columbus's day. The others who might risk exposure seem at first blush to be a surprising group of susceptibles: archaeologists. They might—though this is even more of a long shot than biological warfare—be exposed to smallpox via the mummies and other artifacts they uncover. In 1985, Dr. P. D. Meers of the Public Health Laboratory in London wrote a letter to *The Lancet* suggesting that archaeologists might be "prudent" to get small-pox vaccinations. The variola virus is remarkably hardy, he wrote, and has been shown to survive in cool, dry conditions for many years. He referred to two German physicians who collected scabs from patients after a 1954 outbreak in Leiden of variola minor, a mild form of small-pox. The doctors placed the scabs inside an envelope, and tucked them away in a cupboard. Every year thereafter, they took out the envelope to see whether any infectious variola still remained, then closed the envelope and put it back in the cupboard. They were still finding active virus inside thirteen years later.

Many crypts maintain atmospheric conditions far more conducive to variola survival than an ordinary cupboard; for this reason, Meers wrote, variola is even more likely to be a problem for archaeologists who are, for instance, uncovering bodies preserved in permafrost.


First the cattle started going wild and dying. A dairy farmer in south-ern England offered the earliest reports, in 1986, of cows exhibiting weird behavior: agitation, fearfulness, aggression, and eventually the inability to stand. Within months other farmers reported the same strange symptoms, and it became clear that some infection was rav-aging the herds of Great Britain and threatening the safety of the na-tion's meat supply.

But what truly captured people's attention—even more than the eventual loss of thousands of cattle to infection or slaughter—was the death in Bristol of a Siamese cat. This was the spring of 1990; soon two more cat deaths were confirmed. Before they died, the cats had gone crazy just the way the cows had. Here was a disease, then, with no respect for species boundaries, capable of felling even the kittens who shared people's flats. Could human cases of "mad cow disease" be far behind?

British epidemiologists do not know yet how mad cow disease will end. But they think they have figured out how it began: with a fall in the price of wax. In many parts of the world, including England, cattle feed is fortified with a protein mixture made from the remains of sheep. For years, scientists have known that some sheep flocks carry a brain disease called scrapie. Like mad cow disease, scrapie is a type of spongiform encephalopathy, so named because it riddles the brain with holes, like a sponge. Other species have their own forms of spongiform encephalopathy, including mink, elk, and humans. The human forms are two: kuru and Creutzfeldt-Jakob disease.

Before 1980, when British wax prices were high, producers would mix sheep carcasses with organic solvents to extract the fat to make wax. Afterward, the protein was rendered for use in animal feed. But when tallow prices fell in England, the solvents were no longer used. This may have allowed the spongiform encephalopathy virus to persist, even through the high-temperature rendering process, and show up in the animals' diets.

The problem has continued into the 1990s. To date, at least eighteen thousand cattle have been deliberately slaughtered and burned, an unknown number of domestic cats have died (cat food also sometimes contains sheep offal), and dozens of zoo animals have become infected by eating virus-laden feed derived originally from sheep. Among the dead zoo animals were two kudus, mother and son, that offered the first bit of evidence that the virus could be transmitted in utero—since the baby never ate infected feed and became sick anyway. That means that not only the infected cattle but also *their* offspring are probably tainted. British schools have removed beef from their lunch menus, and importers in France and elsewhere have blocked the importation of British meat.

Because of the way the mad cow virus seems to persist, the condition may remain a problem in Great Britain for the immediate future. Even though officials have determined what caused the epidemic and have taken steps to stanch its spread, they expect cases of mad cow disease to continue until the mid-1990s. By the time the mad cow threat is truly passed, another hazard of strange animal virus infection may be grabbing headlines instead.

# 5

# *Do Viruses Cause Chronic Disease?*

A barnyard pathogen wouldn't seem to have a lot to teach us about the relationship between viruses and chronic diseases. But the surprising outcome of a poultry vaccination campaign in the 1970s indicates just how much like chickens we really are—or, perhaps, how much unity there is in nature. The vaccine in question was for Marek's disease, caused by an avian (bird) herpesvirus. Chickens with Marek's disease develop lymphoma, a fatal cancer of the lymphatic system, the pathway through which immune cells circulate. During the height of its spread, Marek's disease caused an estimated $100 million to $200 million a year in lost revenues for the poultry industry in the United States alone. But the vaccine, derived from a related turkey virus, successfully controlled the international chicken plague.

What is most notable about Marek's disease is what happened after that vaccine was introduced. The chickens given the Marek vaccine didn't just stay free of lymphoma; they became bigger, fatter, healthier, and better egg layers than unvaccinated chickens. The conclusion: many chickens who had previously appeared to be healthy must have been chronically infected with the Marek's disease virus, but at levels so low it was causing no outward signs of disease. The virus was having a subtle effect, though. It was, in some manner, preventing the chickens from growing and producing to their full potential.

Maybe some people are, like those chickens, working at less than full capacity because of a persistent viral infection. Maybe a chronic state of subclinical infection helps explain why some people are shorter than normal, fatter than normal, sleepier than normal, more dim-witted than normal. And maybe a chronic infection even helps explain the opposite: why some people are taller, slimmer, more alert, or smarter. It would be a strange irony indeed if some of the attributes we most value, the qualities that make us believe we are most human and most individual, are traceable to the subtle workings of a virus. Most of us would rather not even admit to the possibility of this con-nection; who wants to believe that a chance encounter with an invisible microbe accounts for our idiosyncrasies, our strengths and weak-nesses, our very personalities? Somehow it seems more acceptable for those traits to be ascribed instead to a gene. While a gene might be equally invisible, and might even be contaminated with genetic ma-terial from an ancient viral infection, at least it seems to have arisen from something within us, from our personal family lineage, rather than from some random outside invader.

But unsettling though it may be to dwell on the fact, scientists have known for years that lifelong characteristics can indeed be formed after a fluke exposure to the wrong virus at the wrong time. A little girl encounters poliovirus at the age of eight, and forever after one leg is smaller than the other; she will always walk with a limp. A fetus is exposed in utero to the rubella (German measles) virus and at birth is profoundly and permanently deaf. A middle-aged man comes down with a herpes simplex infection that goes to the brain, and although he recovers, his once-brilliant capacity to retain details is never quite the same again. We have come to accept the fact that an acute infection, if it occurs during an especially vulnerable period in a person's life or if it involves an especially pernicious virus, can indeed shape us for the rest of our days. From there it's just a short leap to accepting the fact that a *chronic* infection with a virus might have similar lifelong consequences.

Because viruses can live inside cells for many years, they often co-evolve a peculiar kind of symbiosis, an arrangement through which both host and parasite can continue to survive. The virus persists with-out killing the cell, and the cell tolerates the virus without trying to disgorge it. When such persistence occurs, the virus is essentially hid-den from the immune system. It is not a part of the cell, but none-theless it resides peaceably inside it, with neither the cell nor the organism putting up much of a fight. This means that the traditional signs of an active virus infection—an increase in interleukin and other antigen-fighting proteins known as cytokines, the circulation of specific antibodies, the tagging of infected cells with virus proteins on their surface—are missing. Instead, the only indication that the cell is hosting an alien is that the cell may not quite be performing up to snuff. And that cell's true, uninfected potential might never be attained unless something dramatic happens—something on the order of the Marek's disease vaccine.

The notion that chronic illness is infectious in origin might seem a startlingly new idea, but in truth it is a reassessment of an idea that's very old. "Earlier in this century there was a popular notion that hidden infections were the cause of all sorts of ailments, from headaches to arthritis," writes Richard Krause of the National Institutes of Health. "Indeed, for a time it became popular to remove the large bowel to relieve the body of its putrid eluvium. And teeth were extracted and tonsils removed in an effort to eliminate any possible focus of infection."

What is new, though, is the precision with which scientists can spot infection and link it to a problem. So in the same way that viruses that appear to be "emerging" have really been there, unnoticed, for many years—until a chance encounter with another species, or another region of the world, brings that "new" virus to somebody's attention—persistent viruses that appear to be "emerging" as a cause of chronic disease have not changed; all that has changed is our ability to detect and understand them. In both situations, the apparent emergence is an artifact of growing scientific sophistication. When virologists in the field and in the laboratory can isolate and characterize the morbillivirus that kills seals and dolphins, they can refine their analysis of what would otherwise be called a "mysterious plague" and can confirm that yes, a new marine mammal virus has emerged. Similarly, when scientists can actually find traces of viral particles, or at least their nucleic acid, in bits of tissue taken from chronically ill individuals, an understanding develops of how viruses engage in a kind of mutiny of the very organisms they depend upon for survival—and of how the occult presence of an alien bit of genetic material can have lifelong, chronic effects.

For decades, virologists who wanted to draw a connection between an illness and a virus could do so only by looking for antibodies, which are relatively easy to find, rather than for the elusive virus itself. Antibodies are footprints of a previous viral infection, but usually all they reveal is that an infection *has* occurred at some time in the past; they do not indicate *when*. In general, the existence of antibodies in the bloodstream conveys no information about the temporal or causal relationship between infection and impairment. But now virologists are able to spot actual virus pieces rather than their ghostly antibodies, which has allowed a giant leap forward in the ability to trace infections. Scientists can look for active, infectious viral particles in the tissues of sick people or for viral genes that are dormant but ready to be activated when conditions are right. They can say with conviction that the virus is still present—not that it was present at one time, as is shown by finding antibodies, but that it is active and replicating right at that moment. This turns laboratory virology, which not so long ago resembled an archaeological dig, into a real-time adventure more like a scavenger hunt. With new precision, scientists can now uncover more about the workings of the virus, including its mechanisms for taking up permanent residence right inside the cell. And for the first time, they can begin to understand how viruses are involved in creating lifelong and often incredibly subtle damage. "This insidious mode of viral activity may underlie many human illnesses," writes Michael B. A. Oldstone, head of the laboratory of molecular immunology at the Scripps Clinic in La Jolla, California, "such as certain kinds of growth retardation, diabetes, neuropsychiatric disorders, autoimmune disease and heart disease, [all of which] have not been suspected to have an infectious cause."

An understanding of the infectious nature of chronic disease had eluded scientists forced to depend on antibody detection systems, in large part because persistent infections rarely involve antibodies at all. The way most viruses manage to set up a persistent infection is by evading the immune system entirely; antibodies just don't enter the picture. Luckily, scientists now are capable of looking not just for antibodies, or just the virus particle itself, but also for the virus's DNA or RNA. They use nucleic acid probes that are perfect mirror images of the DNA or RNA of the virus they're looking for. A probe is attached to a substance that will light up under the microscope when activated, so the scientist can see easily whether the virus of interest is there. If there is any viral DNA in the cell, the complementary strand will attract it like a magnet; the probe's nucleotides automatically seek out the nucleotides that are their natural pairs. The probe pairs up with its mated DNA or RNA, resulting in what is known as a hybrid double helix; this technique is called nucleic acid hybridization.

Hybridization is completed by scouting for the hybrids themselves, either right in the cell or on a sheet of special filter paper onto which the nucleic acids have been transferred. The way the hybrids are looked for depends on what type of marker was attached to the original probe. If the marker was a radioactive compound, for instance, the scientist looks under the microscope for regions of radioactivity—that is, regions that are lit up on a darkened field. If the marker was an enzyme, the scientist first adds a solution to the cells that will destroy all molecules except those containing the enzyme—and those are the bits that will show up under the microscope. Sometimes, the probe has not been labeled at all, but the scientist instead uses an antibody that is itself a hybrid—it has a radioactive compound attached. After the hybrids are formed, the scientist adds this tagged antibody, which is expected to attach only to nucleotide strands that contain the virus sequence under investigation. Once again, if the slide lights up under the microscope, the virus is there.

There is one important shortcoming to nucleic acid probes: for the probe to work, scientists must know what to probe *for*. The complementary strand must be constructed of nucleic acids strung together in just the right order, with an A located wherever the virus has a T (or, in the case of RNA viruses, a U), and a C located wherever the virus has a G. To create such a probe, they must begin with a clearly sequenced entity, or at the very least with the virus itself to use as a template. But what if they do not know what virus they are looking for? How could scientists ever find and identify new viruses whose sequences have not even been imagined? In truth, the only way to identify a completely new virus is to isolate it from the host tissue and see it under an electron microscope. This method is expensive and technically quite difficult—and it is especially unlikely to work with persistent viruses. In most persistent virus infections, the cells look perfectly normal from the outside; the virologist, looking at the cell under the microscope, would see no reason to search further for a virus to isolate.

Some newer lab techniques manage to get around this problem. One of the most promising, which also managed to implicate HIV as the cause of several mysterious deaths that predate the AIDS epidemic, is polymerase chain reaction (PCR). Invented in the mid-1980s and widely regarded as one of the most significant developments in molecular biology to date, PCR does have the same drawback as nucleic acid hybridization: it requires that scientists know, at least in general, what virus they are looking for. But because PCR is so exquisitely sensitive to the presence of even one virus particle, virologists can also stumble on to findings that are wholly unexpected.

PCR uses the same basic trick that nucleic acid hybridization does: attracting a viral nucleic acid by introducing a complementary template into the cell. But in PCR, an enzyme is also introduced, as well as millions of free-floating nucleotides, to allow that template actually to manufacture more and more new chains of nucleic acid. So the single version of the original "hybrid" becomes two, then the two copies become four, and the four become eight. In rapid logarithmic progression, the amount of genetic material in a given sample is increased ten billion–fold in a single hour. But this staggering ability to multiply any virus particle—which is what makes PCR so valuable to the virologist looking for a virus that, like HIV, might hit just one in every ten million cells—also is the technique's greatest curse. PCR is as sensitive to laboratory contaminants as it is to whatever agent is being studied; if an alien sequence finds its way into the sample, it is amplified indiscriminately along with everything else. A single stray particle from a technician's finger or a dirty test tube can throw off the results of months' worth of experimental work.

One of the most intriguing theories to explain how viruses cause chronic diseases has been offered by Michael Oldstone of Scripps, who bases it on his notion of the "luxury functions" of a cell. Oldstone's thinking goes like this: Every cell has certain jobs it must perform in order to survive. A cell's basic jobs are to make more of its own DNA, to produce certain proteins to maintain its integrity (proteins, for instance, that keep the cell membrane in good repair), and to manufacture enzymes for cellular metabolism. But every cell also has some functions that are, in a way, luxuries—not really essential to the cell's survival but part of the way in which the cell contributes to the optimal performance of the organism as a whole. An individual cell from a human pituitary gland, for instance, can survive even if it secretes less than the normal amount of growth hormone—as long as it keeps producing enough cellular nucleic acid, cellular proteins, and cellular enzymes. But even though that pituitary cell might be keeping itself alive, it's doing precious little from the vantage point of the organism itself. When enough pituitary cells slack off in the production of growth hormone, the individual cells might still be alive—but the organism might not. Similarly, a pancreas cell can survive even if it secretes less insulin, and a brain cell can manage even if it secretes less acetylcholine. But if the pancreas as a whole manufactures less insulin than normal, or the brain as a whole less acetylcholine, the entire organism suffers. From the point of view of the cell, then, the production of growth hormone, insulin, or acetylcholine is what has come to be known as a luxury function. But from the point of view of the organism, these important chemicals are anything but luxuries; without them, the only option is death.

When a cell's luxury functions are impaired, *cherchez le virus*. Sometimes this is quite difficult to do, because the virus slips into the body totally unseen. If a viral infection occurs during a particularly vulnerable point in life—most often in utero or in the newborn period, when the immune system is not yet fully developed—then the body's surveillance network never goes into action. No natural killer cells are mobilized, no antibodies are made, no T cells are set in motion; indeed, the only indication that an infection is under way is that some of the infected cells are not working quite the way they are supposed to. Oldstone and his colleagues came to this conclusion after years of studying LCMV (lymphocytic choriomeningitis virus), a mouse virus that establishes a chronic, persistent infection in mice infected before or shortly after birth. Significantly, the manifestations of infection differ from one mouse strain to another. In most mouse strains, LCMV goes to the thyroid, interrupting thyroid hormone production. In other strains, it homes in on the brain and creates a deficiency of the brain chemical somatostatin. And sometimes, LCMV

Figure 3. PCR amplifies tiny stretches of DNA by using a temperature-cycler that raises and lowers temperature in a test tube to stimulate particular reactions. First the test tube—containing a sample of unknown DNA (A)—is heated so the double stranded DNA splits apart (B). Then the sample is cooled, and a mixture of "primers"—nucleotide strands complementary to the beginning of the DNA chain being looked for—are added. The primers bind to the beginning of each of the two single strands (C). The temperature is then lowered further and a polymerase enzyme is added to the mixture, along with an enormous number of unattached nucleotides. The polymerase retrieves the nucleotides and brings them to the DNA chain, where they are attached, in order, to their corresponding nucleotides (D). When the entire chain is constructed, there are two complete copies of the targeted sequence (E). The temperature is then raised again to split apart the two chains, and each one builds two new copies. In this way, two copies become four, four become eight, and after one hour, about 10 billion copies of the target sequence are in the test tube.

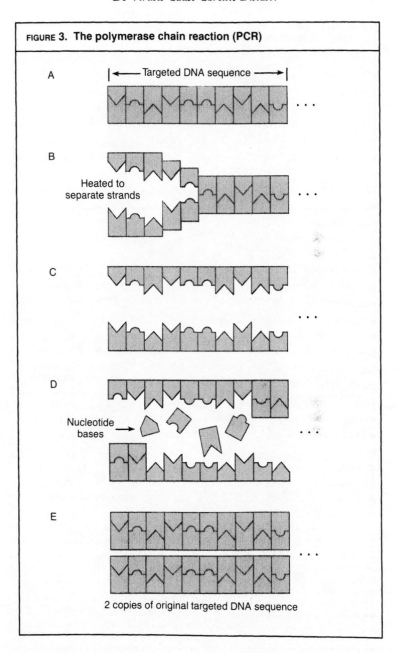

FIGURE 3. The polymerase chain reaction (PCR)

A — Targeted DNA sequence

B — Heated to separate strands

C

D — Nucleotide bases

E — 2 copies of original targeted DNA sequence

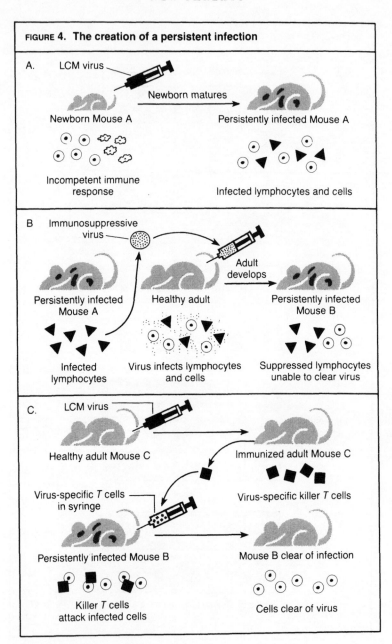

FIGURE 4. The creation of a persistent infection

A. LCM virus

Newborn Mouse A → Newborn matures → Persistently infected Mouse A

Incompetent immune response

Infected lymphocytes and cells

B Immunosuppressive virus

Persistently infected Mouse A

Healthy adult → Adult develops → Persistently infected Mouse B

Infected lymphocytes

Virus infects lymphocytes and cells

Suppressed lymphocytes unable to clear virus

C. LCM virus

Healthy adult Mouse C → Immunized adult Mouse C

Virus-specific killer T cells

Virus-specific T cells in syringe

Persistently infected Mouse B → Mouse B clear of infection

Killer T cells attack infected cells

Cells clear of virus

heads for the pituitary gland and impairs growth hormone production, leading to abnormally small size and abnormal metabolism of glucose (sugar).

The first mice Oldstone studied were dwarf mice with congenital LCMV infections in the pituitary gland, the pea-size organ at the base of the brain that directs the production and secretion of most of the body's hormones. "When we examined the infected [pituitary] cells under the microscope, we saw no evidence of cell injury or inflammation," Oldstone recalls. If pituitary cells looked normal, he wondered, why weren't they behaving normally? Somehow, without changing the viability of the cells themselves, the virus had "disturbed homeostasis and caused disease."

The next important question was how LCMV could slip by all lines of defense and insinuate itself into a particular type of mouse cell. And it is in answering this question that Oldstone devised his theory of how other viruses can set the stage for chronic human conditions, among them hypertension, atherosclerosis, diabetes, short stature, and mental illness. In Oldstone's view, many of these problems can be traced to the type of persistent viral infection that can occur only when a virus manages to derail some immune system cells. Looking at LCMV and mice again, he concluded that at first the virus could cause persistent infection only in newborn mice, which have an incomplete immune response. All he needed to do was inject LCMV into the newborn, and the animal was infected for life. But here is where it gets even more interesting—and worrisome: during that persistent infection, something happened to the LCMV. Mutants of LCMV arose that were capable of quelling *any* immune response against them, even by immune-competent animals. These mutant LCMV were able to do something that their normal counterparts could not: set up residence in the lymphocytes, the very cells that are intended to fend off viral infection. In the unusual environment of an immunosuppressed organism, LCMV variants that were able to live in lymphocytes thrived,

Figure 4.

A. Newborn mice infected with LCM virus cannot mount an immune response; they develop a persistent LCM infection in the brain, spleen, liver, and kidney. Some lymphocytes in these mice are also chronically infected.

B. Virus from these LCM-infected lymphocytes has changed sufficiently to suppress lymphocyte functioning even in healthy adult mice, allowing LCM to establish a persistent infection.

C. These persistently infected mice could be cured with injections of killer T cells taken from healthy mice that were also exposed to LCM virus. These other mice make virus-specific killer T cells, which attach to LCM-infected lymphocytes and clear them from the mouse.

a bit like burglars getting a special kick out of raiding the refrigerator of their absent "hosts."

When LCMV was isolated from persistently infected lymphocytes and injected into healthy adult mice, the change in the virus became obvious. Unlike the original LCMV, these mutants—which arose and proliferated in the neonatally infected mice—were able to invade the lymphocytes even of healthy mice, and thus could establish a persistent infection in *any* mouse. A virus that had set up a persistent infection in immune-deficient mice had changed sufficiently to be able to set up a persistent infection in mice with perfectly functioning immune systems.

Could the same thing happen with human viruses? The possibilities are quite frightening. We now have on this earth a highly inflated number of people without functional immune systems: AIDS patients kept alive for years with antiviral drugs; transplant recipients who take immunosuppressing drugs for their entire lifetimes so they don't reject their new organs; people with autoimmune diseases who are also treated with immunosuppressants, in the hopes that turning off the immune system altogether will effectively stop the self-destruction. In each of these categories, as we have seen, immune suppression places these people at great risk of illness from otherwise benign endogenous viruses that had been quelled while they were healthy. But Oldstone's work gives us reason to be worried that these viruses are a threat not only to the immunosuppressed individuals but also to others who are quite healthy. Is it possible that immunosuppression constitutes a new environment, comparable to the environment of LCMV-infected newborn mice, in which mutant viruses capable of invading lymphocytes will be able to proliferate? If so, even viruses that currently are not capable of establishing persistent infections may, eventually, evolve the ability to evade a functioning immune system. The existence of all those immunosuppressed people might lead at some point to the evolution of some new herpeslike viruses that can set up lifelong residence and cause permanent, subtle damage to the peak performance of chronically infected human beings.

This brings us to another characteristic of viruses, which can also explain their chronic effect: their tendency to seek out and invade specific cells that will provide a comfortable home. This is the quality known as tropism. Many persistent viruses readily hide in only one type of cell; the cell is ideally suited to their total isolation from the cells of the immune system. Tropism results from a perfect fit between the coat of the virus and the receptors of the invaded cell; usually, the target cell grants entry to the virus because it looks so much like a substance the cell ordinarily would *need* to let inside, like an important nutrient. LCMV, of course, is tropic in certain strains of mice for the pituitary gland, and in other strains is tropic for the thyroid, pancreas, or brain. Many human viruses have equally well-defined tropisms. Influenza virus is tropic for the respiratory tract, rabies virus is tropic for the brain, hepatitis virus is tropic for the liver. When hepatitis B virus sets up a persistent low-grade infection in its preferred organ, the result can be, for the infected individual, a lifetime of chronic liver disease. Another virus, known as coxsackievirus (named after the town in upstate New York where it was first isolated), is in some individuals quite tropic for the heart. In fact, persistent infection with this highly contagious virus has been associated with a serious condition known as cardiomyopathy (also called myocarditis), a heart enlargement that affects babies and young children, often after they have recovered from a bout of a stomach virus that resembles influenza. Most children with heart involvement eventually recover, but many develop progressive congestive heart failure and will ultimately need a heart transplant. The relationship between cardiomyopathy and coxsackievirus was clarified in the late 1980s, when scientists in London and Munich examined more than seventy cardiomyopathy patients. In roughly half, the heart muscle contained RNA sequences from coxsackievirus. In a comparable group of forty other patients, who also had cardiac disease but not cardiomyopathy, no coxsackievirus RNA could be found.

One notorious virus that is highly tropic for a particular region and is able to establish a long-term persistent infection without killing the infected cells is Borna virus. It heads straight for the brain, particularly the limbic system, the brain region where the full array of emotions—rage and astonishment, lust and euphoria, passion and pain and despair—originates. Borna virus is the bizarre psychoactive virus that was manipulated by mad scientists for use in germ warfare in the science-fiction movie *Warning Sign*. It was no coincidence that Borna virus had the dubious honor of that offstage role. By general consensus, it is one of the strangest viruses known.

It was in the pastures near the Saxon city of Borna that the odd behavior of some horses and sheep was first remarked upon. The year was 1813, a century and a half before mad cows were discovered in England. The animals became rigid and frenzied in their motions; a large number died of encephalitis. Although nearly 150 years passed before the exact cause of the disorder, named Borna disease, was identified, investigators along the way were able to use classic virology techniques to show that it was caused by a virus rather than by some other pathogen. They used the method developed by Martinus Beijerinck at the turn of the century, when he first inferred the existence of the infectious agents he was eventually to call viruses. A scientist took samples of the brain of an animal with Borna disease, ground up the tissue, passed it through a filter with pores small enough to screen out all known pathogens, and injected the resulting liquid into the brains of laboratory animals. The lab animals soon came down with many of the same symptoms of the animals naturally infected—confirmation that something infectious remained in the brain mixture even after bacteria and other organisms were eliminated. The only thing that could have passed through the filters was a virus.

That Borna disease soup, injected directly into the brains of laboratory animals, was able to infect an unusually wide range of species: mice, rats, hamsters, guinea pigs, rabbits, tree shrews, birds, and monkeys all could get Borna disease experimentally. Interestingly, different animals exhibit slightly different symptoms, including many antisocial behaviors not usually associated with infection. Rats, for instance, become extremely aggressive immediately after infection and within a few weeks become extremely lethargic; about half of them overeat compulsively and balloon to two and a half times normal weight before they die. (This, by the way, is not the first time a virus infection has been linked to morbid obesity; mice have been known to become extremely fat after being inoculated with canine distemper virus.) And on certain measures of learning, such as running through mazes, rats persistently infected with Borna virus—who otherwise appear perfectly normal—show definite learning defects.

Tree shrews infected in the lab and housed together in cages exhibit social pathology, of a sort not observed in infected tree shrews housed alone. They become aggressive, repetitive in their behavior, and display altered sexual behaviors like males trying to mate with males. And the behavior of infected rhesus monkeys looks a lot like human manic depression. First there is a manic phase, when the monkeys are aggressive, hyperactive, and ataxic (unable to contract the voluntary muscles at will). Then follows a depressed phase; the monkeys refuse to eat or drink, their limbs become paralyzed, and they seem, as one group of investigators puts it, "apathetic."

Because of the strange behavioral symptoms associated with Borna virus in animals, researchers have naturally wondered whether the virus has any similar effect on human beings. The possibilities are dramatic: maybe the virus can cause problems such as depression, obsessive-compulsive disorder, learning disabilities, hyperactivity, eating disorders, or some of the other human correlates of the animal behavior that has been observed. In 1985, a group of scientists working collaboratively in the United States and West Germany found antibodies to Borna virus in a small proportion of patients with mood disorders, especially manic depression. They found antibodies in 16 patients—12 in Philadelphia, and 4 in Würzburg and Giessen—out of 979 psychiatric patients whose blood was examined. Although this might seem like a minuscule proportion—just over 1.6 percent—it was significant compared with the controls. Of the 200 normal individuals whose blood was also sampled, in both Germany and the United States, none had antibodies to Borna virus.

Of course, the presence of antibodies reveals nothing in itself about the temporal association between infection and illness; those sixteen patients might have been exposed to a Borna-like virus early in childhood or might have been exposed just months before the development of their psychiatric depressions. But even though so few patients had Borna antibodies, virologists still considered the findings to be suggestive of an association. After all, Borna virus is not something that is just floating around; not a single one of the healthy people tested had ever been exposed. Indeed, even among animals, Borna disease is a rarity; the condition strikes herd animals in Europe only sporadically, and has never occurred naturally among any animals in the United States. So those sixteen antibody-positive patients were really something to notice. But the report of their existence, dutifully recorded in *Science* magazine in 1985, just remained in the literature, unconfirmed but intriguing, for years.

Not until 1991 did a group of scientists from Baltimore, Maryland, manage to establish an even clearer connection between Borna virus antibodies and another devastating mental illness, schizophrenia. Kathryn Carbone, an infectious disease specialist at Johns Hopkins

University, and Royce W. ("Bill") Waldrip II, a psychiatrist at the University of Maryland, compared the blood of forty-nine schizophrenics with that of eighteen carefully screened controls. In about 17 percent of the schizophrenics, Carbone and Waldrip found what they were looking for—Borna virus antibodies. And in the control group, only about 3 percent had Borna antibodies, which was a statistically significant difference. The schizophrenics with Borna virus antibodies, who were still in a distinct minority among all schizophrenics in the study, tended to be sicker than the schizophrenics without antibodies. They were more likely to be male, to be poor, and to have chronic (rather than acute) schizophrenia. But still the results are inconclusive, because once again they involve a finding of antibodies rather than the virus itself. Carbone and Waldrip are now trying to isolate actual active Borna virus from the tissues of those schizophrenics who are willing to undergo brain biopsies.

Borna disease virus has two unusual features: it heads directly for brain cells, particularly those in the limbic system; and it gets into the nucleus of the brain cell, where it can take up residence for many years and keep producing more viruses. No other RNA virus is capable of replicating in the nucleus of an infected cell. "It looks to me as though Borna virus is one of a whole new class of infectious agents," says W. Ian Lipkin, a neurologist turned molecular biologist trained in Oldstone's lab at the Scripps Clinic, who with his colleagues was the first to isolate and sequence the Borna disease virus in 1990. "I think it will help explain, and eventually treat, a wide variety of neurological and psychiatric illnesses." It's radical enough to suggest that a virus might *explain* mental illness. But that a virus might help *treat* it? This is a far more dramatic idea. Lipkin wants to use Borna virus as a kind of guided missile, to which he can attach drugs or chemicals the brain needs for optimum functioning. Getting the right compound to the right spot in the brain is a perennial problem for neuroscientists, and Borna virus might hold the key to its solution.

The brain is traditionally a sacrosanct place, held at a remove from the rest of the body by something poetically called the blood-brain barrier. This barrier, though not quite so impervious as the name implies—plenty of neurotoxins circulating in our bloodstream manage to find their way to our brain—has foiled recent attempts to correct brain diseases caused by a deficiency of certain neurotransmitters, the chemicals that control brain functioning. Scientists have discovered which neurotransmitters account for sadness and happiness, for hunger and satiety, for aggression and passivity, but they have been unable to do much clinically with this information because of the blood-brain barrier. Lipkin hopes Borna virus will change that. His goal is to genetically engineer Borna virus so that the genes that make it infectious are eliminated, thus ending up with a Borna virus shell that retains just one essential feature of the original: its ability to get inside a brain cell. Onto this shell, or "vector," he can then attach a gene for a neurotransmitter—say, somatostatin—that an individual cannot manufacture alone. The gene-enhanced vector will then go right into the neurons of the limbic system, and the new somatostatin gene will enable the brain to produce the chemical. This would be much more preferable than the current approach of trying to flood the brain with somatostatin to increase the odds that some of it will find its way to the region where it can do the most good. "Instead of hitting a person with a bomb, in the hopes that you will get up-regulation [increased neurotransmitter production] everywhere, now you can surgically do whatever you want to do," Lipkin says. With Borna virus as a vector, neuroscientists can be far more precise about hitting their targets.

Another unique characteristic of Borna virus infection is that the virus does virtually no damage to the cell, even after it has infiltrated the nucleus. The symptoms seem to be caused by something quite different from simple cell impairment or destruction: they are generated by the body's own immune response to the virus's presence. Borna virus is one of the most dramatic examples of a virus's causing its damage by immunopathogenesis (immune system–created symptoms) rather than cytotoxicity (direct killing of the cell). When adult rats are experimentally infected with Borna virus, they show brain abnormalities (encephalopathy) and behavioral disturbances. But when the rats are made tolerant of the virus, so that they will fail to mount an immune response against it, and *then* are infected with Borna virus, nothing happens. All the brain damage of Borna virus seems to come from the inflammation that results from the body's defenses, not from any cytotoxic action of the virus itself.


In some persistent infections, then, the immune system is in such high gear that it unwittingly turns a rather benign virus into a killer. Something similar might be taking place in people with so-called chronic fatigue syndrome, which remains one of the greatest enigmas in contemporary virology. First recognized as a new condition in 1984, the disorder was for a long while disparagingly referred to as the yuppie flu. Its victims, mostly youngish, ambitious, successful, and female, recoiled at the suggestion that they invented the syndrome because it was their only excuse to slow down. But the psychosomatic label persisted, in large part because the symptoms were so subjective: profound fatigue, muscle aches, headaches, depression, forgetfulness, poor concentration, and inability to get on with the business of daily life. More than a few observers also questioned how these patients— whose chief complaint is bone-deep exhaustion—managed to summon enough energy to form their countless support groups around the country and to fire off dozens of faxes and press releases urging the government to allocate more money for research.

But the accumulation of objective measures that differentiate patients with chronic fatigue syndrome has bolstered the victims' insistence that something physical is going on. Unfortunately, as with so many other conditions, this syndrome probably has many different causes—and, therefore, no single finding that will yield a nice, neat blood test and a simple yes-or-no diagnosis. Some scientists have found that chronic fatigue syndrome patients have a higher-than-normal level of immune system activity, as measured by serum levels of killer T cells. It's as though their immune defenses are stuck in the On mode, perhaps because of past exposure to some virus that is no longer a threat. The body is using all its resources to fight a phantom, a little like what happens to allergy victims, whose bodies mistakenly react to harmless substances like cat dander or ragweed by mobilizing the chemicals usually required to rout out dangerous invaders. With the defenses continually on, the patient suffers all the side effects usually associated with fighting off an active infection—side effects, like achiness and exhaustion, that can be traced directly to the actions of the cell-killing cells called cytokines, the release of which is controlled by killer T cells.

Another component of the immune system, the natural killer cells, are also operating differently in chronic fatigue syndrome patients, but the cells' behavior is paradoxical. In some patients the natural killer cells are working far more aggressively than normal; in others, they are barely working at all. "Think of the immune system as a person who is overwrought," says Anthony Komaroff, chief of general internal medicine at the Brigham and Women's Hospital in Boston. "Just as can happen with people, immune system cells that are overwrought may become ineffectual."

The reason for all this misplaced activity may well be the reactivation of a chronic viral infection. Many human viruses, especially herpesviruses, are capable of slipping into a latent mode for many years. After being beaten back by the immune system following an initial infection, the virus takes up residence in particular cells in the body—liver cells, nerve cells, immune cells, or others, depending on the type of virus involved. Herpes simplex type 1 and type 2 viruses, as well as varicella-zoster, retreat to the nerve cell ganglion, a small collection of nerve tissue lying outside the brain and spinal cord; the Epstein-Barr virus resides in the immune system's B cells; human herpesvirus 6 (HHV-6) and cytomegalovirus also head for immune system cells, though probably the T cells. (All of these examples are from members of the herpesvirus family.) For years, the latent virus remains quiet. Its DNA is there in the cell cytoplasm, but it engages in no active effort to utilize cellular mechanisms to replicate. No active effort, that is, until the immune environment changes in response to some external event. Then the latent virus reactivates, with symptoms that are often far more dramatic than they were when the virus first appeared.

In chronic fatigue syndrome, something happened to the immune system to reactivate one of these viruses. That initiating factor—as well as the precise viruses that are reactivated—is what now eludes investigators. Komaroff thinks that, at least in some cases, it might be emotional stress, which is known to suppress the immune system. He hastens to add, though, that some other factor—such as exposure to an environmental toxin or maybe even another infection—might also derange the immune system enough to let loose the infectivity of a latent virus. After this primary event, and after the virus begins replicating, the immune system shifts gear once again. Now there is an active viral infection to fight. But in this new, vigorous counterattack, something goes wrong, primarily because the viruses being fought are such masters of subterfuge and camouflage. It's a bit like fighting the guerrilla warriors of Vietnam with the bombers of World War II. "There's nothing the immune system can do to get rid of a chronic infection that cannot be eradicated," says Komaroff. "So you end up with a chronic war. And eventually, as with any chronic war, some of the troops will become exhausted."

The scientific troops are becoming somewhat exhausted too. To rally a new enthusiasm for this continually vexing research, the small chronic fatigue task force of the Centers for Disease Control was host of an informal meeting in September 1991. On the agenda was discussion of the three viruses thought most likely to be involved in initiating this weird guerrilla warfare: a human foamy virus, which is a retrovirus not previously thought to cause any human disease; another retrovirus, newly identified in 1990 and thought to be similar, but not identical, to the cancer-causing retrovirus HTLV-II (human T-cell lymphotropic virus); and HHV-6, a virus discovered in 1986. Although HHV-6 at first was associated only with roseola, a mild childhood disorder that causes little more than a fleeting rash, new evidence suggests it may underlie as many as 14 percent of cases of high fever in babies that previously were misdiagnosed as bacterial sepsis or meningitis. British studies also suggest that one of the enteroviruses, a large viral group that comprises three types—poliovirus, echovirus, and coxsackievirus—might also be involved in touching off chronic fatigue syndrome, primarily because they are so easily transmitted, so widespread, and have the same proclivity as herpesviruses and retroviruses to revert to a latent, noninfectious state for many years.

Not so long ago, the British virologist Robin A. Weiss, a scientist at the Institute of Cancer Research in London, called one of these candidate viruses "a virus in search of a disease." He was talking about the human foamy virus—but he could have been describing any one of the others. The first human foamy virus (also called spuma virus) was detected in the early 1970s, more than twenty years after foamy viruses were found in monkeys, chimpanzees, cats, cattle, and hamsters. But even now, no one really knows what foamy virus does. Because it is a retrovirus, its presence in certain cells studied in the laboratory could be a contamination, or it could reflect an ancient infection in the host animal's cells that causes no disease at all. Indeed, no one can even say with certainty whether there really is such a thing as a specifically human foamy virus, rather than a simian foamy virus that occasionally gets into people. The only connection ever made between the virus and human illness was in Czechoslovakia, when a group of virologists from Bratislava found foamy virus isolates in eight of the twenty-eight patients they were examining for a transient thyroid inflammation known as subacute thyroiditis de Quervain.

Like foamy virus, HHV-6 is also a virus in search of a disease. It has been known only for a few years, and the general feeling is that it must have a significant human health effect beyond roseola. And like these two viruses, the new retrovirus connected to chronic fatigue syndrome in 1990 has no known pattern of disease transmission. The scientist who found indirect evidence of this new retrovirus, Elaine DeFreitas of the Wistar Institute in Philadelphia, appeared at a press conference sponsored by the national chronic fatigue syndrome support group just after reporting her findings at a neuropathology conference in Japan. She was cautious in her conclusions, emphasizing that she had merely found a gene sequence that looks like HTLV-II in the blood of a majority of a small group of chronic fatigue syndrome patients. (Thirty-one adults and children were examined; 77 percent of them had the HTLV-II sequence.) "We have never said that this virus or viral gene causes [chronic fatigue syndrome]," she told reporters. "We are just reporting an association between a relatively rare virus" and a relatively rare disease. Even her collaborator, Paul Cheney, the North Carolina physician who provided serum from his patients and who is credited with recognizing an early outbreak of chronic fatigue syndrome in Nevada in 1984, urged care in making any unwarranted leaps about the relationship between the retrovirus and the illness. "It's often a long time between when you have in your hand a virus," he said at the press conference, "and when you have enough studies under your belt [that you can], with some reasonable degree of certainty, claim causality."

The difficulty in proving cause and effect between a virus and a chronic condition is nowhere more obvious than in the relationship between cytomegalovirus and atherosclerosis. Atherosclerosis, sometimes nicknamed "hardening of the arteries," is the growth inside the blood vessels of clogging plaques, which begin as elevated fatty streaks along the blood vessels' inner layer, or endothelium. Over the years, the streaks collect a variety of other cells—immune system cells, smooth muscle cells, fats, cholesterol, blood cells, calcium, and foam cells—and eventually become elevated masses that interfere with blood flow through the vessel. The process is thought to be touched off by one of three things: destruction of the endothelial cells, a change in their metabolism, or a cancerlike transformation of these cells that leads to unregulated growth. Any one of these three things can be initiated by a virus.

The possibility of a viral origin for this common disease was first proposed back in the barnyard, with the Marek's disease virus. In 1978, scientists at Cornell University raised 130 chickens in a germ-free environment, and then exposed half of them to Marek's disease virus by injecting it directly into their bloodstreams. They also designed the experiment so that about half of the inoculated chickens and half of the germ-free chickens were kept on a low-cholesterol diet, and the other half of each group was given cholesterol supplements. All of the chickens that remained germ-free were healthy, no matter what they ate. But of the forty-two chickens that were infected with Marek's diease virus, seven developed atherosclerotic plaques: three of those on a low-cholesterol diet, four of those fed extra cholesterol. "Their blood vessels looked exactly like what you see in human beings with atherosclerosis," says Joseph L. Melnick, a medical virologist at the Baylor College of Medicine in Houston. "If you took slides of their plaques and put them side by side with human plaques, even a pathologist wouldn't be able to tell which was which."

Of course, this set physicians to wondering whether human atherosclerosis might also be caused by a virus. They began by looking for the human corollary to Marek's disease virus. The family of herpesviruses, of which the Marek's disease virus is a member, is highly adapted to one particular host species; a human herpesvirus almost never infects an animal, and vice versa. (A prominent exception to this rule, of course, occurs during occasional occupational exposure, as when lab workers become infected with the monkey herpesvirus herpes B; the result of cross-species herpes infection is almost always devastating.) What herpesviruses have in common, no matter the species involved, is that they can persist for decades without causing any symptoms. Indeed, their perpetuation depends on a cycle of latency followed by reactivation, since it is only during the relatively short periods of reactivation that they are contagious at all.

Each of the seven human herpesviruses, like their animal counterparts, takes up residence in a particular type of cell and remains there in a latent state, only to erupt in occasional flare-ups when the individual's immune defenses are down. Herpes simplex type 1 causes cold sores, and lives in nerve cells when not making itself visible as unsightly "fever blisters" around the mouth and nose during times of illness or stress. Herpes simplex type 2 causes genital sores, and it reverts to its latent state in the nerve cell ganglia. Herpes simplex viruses are infectious only during reactivation. Type 1 can be passed by a kiss, type 2 by sexual intercourse or from mother to baby during childbirth.

Varicella-zoster virus, another herpesvirus, has a two-stage pattern of infection. When an individual is first exposed, usually during childhood, the result is chickenpox (varicella). Then the virus goes into retreat in the nerve cells and stays there for nearly a lifetime. Fifty or sixty years later, when the host's immune system is less vigilant, the virus has an opportunity to re-emerge in a new form: as herpes zoster, or shingles. This is a painful skin eruption that originates in the nerve cells where the varicella-zoster virus had been in residence all those years. An individual in the first half of a shingles attack, which usually lasts about a week, is highly contagious, and someone who never had chickenpox is likely to catch the herpes-zoster virus by getting too close. Other human herpesviruses are the Epstein-Barr virus, which causes mononucleosis and a rare form of cancer known as Burkitt's lymphoma; the two most recently discovered members of this family, human herpesvirus 6 and human herpesvirus 7; and cytomegalovirus, which causes a mild flulike illness, significant birth defects in infected fetuses and some forms of mononucleosis, and possibly contributes to atherosclerosis.

Cytomegalovirus, familiarly known as CMV, is the human herpesvirus that seems most likely to behave in people the way Marek's disease virus does in chickens. In several recent studies, cytomegalovirus DNA has been isolated from tissue taken from the arteries of people with atherosclerosis. And in heart transplant recipients, the presence of actively replicating CMV has been shown to accelerate the development of atherosclerosis in the new heart's vessels, leading to total blockage in a matter of months—rather than the decades it ordinarily might take. At the Stanford University School of Medicine, for instance, researchers found that heart transplant patients with CMV infection were twice as likely to be dead five years after the transplant as were heart transplant patients who were free of CMV. After ten years, infected patients were three times as likely to be dead. And in one study conducted at the University of Limburg in the Netherlands, artery samples were compared in two groups of older men: forty-four who were undergoing coronary bypass surgery, and thirty-four who had died of diseases not associated with atherosclerosis. Using PCR, the technique that amplifies even trace amounts of genetic material, the investigators found DNA from cytomegalovirus in 90 percent of the men with atherosclerosis, compared with just 53 percent of the men in the autopsy control group—a statistically significant difference. Similar findings have been made at the Baylor College of Medicine.

The evidence in support of the theory that CMV causes atherosclerosis is "all circumstantial," says Melnick of Baylor, but "very much worth pursuing." It's at least compelling enough to make Melnick think that a CMV vaccine could significantly reduce the burden of cardiovascular disease in the Western world. "They did it with chickens, at least experimentally," he says. "They took the Marek's disease virus, inactivated it with a chemical just the way you do for the Salk polio vaccine, inoculated chickens with it, and then a few weeks later they inoculated the chickens with active Marek's disease virus. Those chickens never got atherosclerosis." Even chickens fed high-cholesterol diets, once they were protected with the avian Marek's disease vaccine, failed to develop plaques in their arteries.

But people are not chickens, and CMV—unlike Marek's disease virus—has not been associated with any major acute disease. It just doesn't seem to be an especially dangerous virus in terms of either human agony or economic pain. Most people who are exposed to CMV during childhood or adolescence never even know they were sick; more than half of all thirty-five-year-olds in the United States have CMV antibodies without ever having suffered. (One major exception is when CMV infects a fetus; the virus is probably the single biggest cause of congenital mental retardation and learning deficits in the United States.) Unless a clear relationship can be established between CMV and atherosclerosis, getting public or private funding to disseminate a CMV vaccine could be as difficult as getting funding for a vaccine against the common cold.

The second leading cause of heart disease death in the Western world, after atherosclerosis, is high blood pressure—another chronic condition that has been associated with viral infection. "It's been a difficult theory for anyone to buy off on—that there might be an infectious origin for some well-known chronic diseases," says epidemiologist James LeDuc. But he thinks that many cases of high blood pressure in the United States can be traced to a chronic viral infection, and he thinks he even knows the virus involved: a rodent-borne Asian virus from the Hantavirus family. "The Hantaviruses represent a prime example of an emerging virus disease which offers both new challenges and perhaps explains longstanding questions," he has written. The most significant long-standing question answered by the virus is one of great social import: Why do people living in poverty, especially black people, experience such a high rate of hypertension, stroke, and kidney disease? LeDuc thinks it may be, at least in part, because they are chronically infected with a virus carried by the alley rats with whom they often share their homes.

LeDuc's theory has its roots in the work of Ho Wang Lee, the South Korean biologist who discovered in 1980 that the rats of Seoul were harboring a type of Hantavirus. Lee's finding that urban rats carried what came to be known as Seoul virus meant that a huge number of people were exposed to it. The rat, after all, lived in crowded, squalid urban areas, with people packed upon people, in every major city on earth. In the early 1980s, medical detectives set out to look for Hantavirus antibodies in the blood of rats in cities in Europe, Scandinavia, Africa, and South America. In the United States, LeDuc and his colleagues chose a handful of port cities and began to catch and take blood samples from the rats hanging around the wharfs. The vast majority had antibodies to what they called a "Seoul-like" virus, a member of the Hantavirus family but not exactly identical to the one found in Korea. By the time they had reached rat old age, more than 70 percent of American rats had been exposed to the Seoul-like virus. Everywhere LeDuc's team looked, there was Hantavirus. Everywhere their international collaborators looked, there was Hantavirus. In Asia, in Europe, in South America, in port cities and in cities far from shore, there was Hantavirus. In the United States alone, Hantavirus was found in rats in Baltimore, Honolulu, Houston, New Orleans, New York, Philadelphia, San Francisco, and even in the inland city of Columbus, Ohio. "What we were observing was not a recent introduction or a transformation of this wild Asian virus into a new urban virus," LeDuc says. "These were two independent viruses, and the urban virus had been well established around the world for a long time."

LeDuc engaged in plenty of back-breaking field work in the alleys and trash cans of America's inner cities, especially the city of Baltimore, Maryland, which is less than an hour's drive from his home base at Fort Detrick. Using skills developed during his years as a mammalogist in Africa, LeDuc and his team would set rattraps in the alleys behind Baltimore's shabby downtown row houses, a short walk from the city's glittering retail showcase, Harborplace. When they caught a rat, they would carefully pin the rat to the side of the trap and inoculate it with an anesthetic. "We would take it away from where most people were, and get to work," LeDuc recalls. "First we weighed it as a rough measure of how old it was; in general, the older the rat, the more it weighs. Then we inspected it for scar tissue to indicate how many fights it had been in. Again, the older the rat, the more scars it has." After that, the scientists cut a slit in the rat's tail and collected its blood in a test tube. Then they killed it and disposed of it. Sometimes, they also took tissue samples to study at the lab. And all the while, the men tried to exercise caution in their rough-and-ready field lab, knowing that the virus they were interested in was highly contagious. They shunned whole-body "space suits"—they didn't want to look too conspicuous—but they did wear masks and gloves. And, as LeDuc puts it, "we tried to work with the wind at our backs, so any virus we were generating would be blown away from us."

Once Hantavirus antibodies were found in rodents throughout the world, the next step was to look for evidence of Hantavirus infection in human beings. The problems of finding rats to take blood from turned out to be nothing in comparison with trying to find humans to bleed. LeDuc teamed up with James Childs, a scientist who had spent a year working in the army lab and had then moved to Johns Hopkins University in Baltimore. "It took a couple of years to establish enough rapport with the physicians in the hospital to get access to the correct patient population," LeDuc says. By 1989, the two men had screened the blood and urine of eleven hundred hospitalized patients at Johns Hopkins and had found Hantavirus antibodies in fifteen. When they checked the blood pressures of these fifteen people and compared them to people of the same race, sex, age, and socioeconomic status, they found that those with signs of viral exposure were significantly more likely to have high blood pressure and chronic kidney disease. They also examined the blood of four hundred patients at a

Baltimore dialysis center and found antibodies to Hantavirus in eight of them—nearly double the rate of the hospitalized patients.

This, for the first time, was evidence that those antibody-positive rats clambering about in the back alleys had indeed passed on their infection to at least some residents of Baltimore. If the U.S. strain of Hantavirus behaves like the Asian strains, it is shed in the animals' urine and feces and can be transmitted through soiled material or possibly through the air if the virus dries and becomes airborne as tiny infectious particles. Another possible route of infection is via alley cats, which are even more likely than rats to come in direct contact with people. In China, Hantavirus-infected cats have been identified, though scientists still do not know whether they are capable of growing the virus in their intestines and shedding it in their feces the way rodents do. LeDuc would like to explore this connection, but he thinks conducting research on cats would be too politically sensitive just now. Childs has already collected dozens of samples of cat blood from Baltimore streets as part of his dissertation research, but he decided against experimenting on them for fear of offending animal welfare activists, whose antivivisectionist actions have grown especially militant in recent years. For now, no one is investigating whether American pet cats harbor Seoul-like virus.

Another way in which politics has tainted this line of research has been an even greater frustration for James LeDuc. Sitting in a vault in an army warehouse, stored in six hundred steel flasks, are freeze-dried samples of the blood of about 240 young men who came down with Korean hemorrhagic fever during the Korean conflict nearly forty years ago. (Korean hemorrhagic fever is caused by exposure to the Asian strain of Hantavirus; it hardly ever occurs naturally outside the Orient, even in the cities in which Hantavirus has been found in rats.) Someone knows where these men live today, and it should be an easy matter to track them down and see if they were more likely to develop kidney failure or hypertension than a group of Korean vets who did not catch hemorrhagic fever. It's a beautiful natural experiment, an easy way of establishing the chronic effects of a long-term virus infection. But LeDuc cannot convince his superiors that he should spend his time on a follow-up study. "Inner city hypertension," he says, "is not considered mainstream military medicine."

Not military, perhaps, but certainly mainstream. High blood pressure is one of the most common diagnoses in the United States today, affecting more than 35 million Americans (more than 20 percent of all adults), and a full 85 to 90 percent of those people have hypertension for which no cause can be found. Kidney and urologic diseases cost the nation more than $50 billion a year in medical care, including drugs, hospitalization, and dialysis. If even as small a portion as 2 percent can be traced back to Hantavirus—the proportion of patients in the Baltimore dialysis centers found to have Hantavirus antibodies— then the nation is spending some $100 million every year on a condition that could be fully prevented through rat-control methods or, ultimately, a Hantavirus vaccine. As with CMV and atherosclerosis, the notion that a vaccine could prevent a major chronic disease is an exciting one. But with scanty evidence about cause and effect, and with so many infectious diseases of immediate consequence to worry about, scientists and drug companies alike are rather uninterested in devoting the time and money it would take to bring such a vaccine to market.

# 6

# *Tropical Punch:*
# *The Lethal Arboviruses*

For the seven-year-old Massachusetts boy, the summer of 1990 had started out ordinarily enough. He spent a lot of time playing around the house, in the town of Marshfield, and he also went camping near Plymouth. But in mid-August he began to feel achy and feverish, and he stayed in bed for a week. When his temperature reached 105° F (40.5° C) and he began to have seizures, his frantic parents took him to the emergency room.

The boy, it turned out, was the state's first case in six years of eastern equine encephalitis (EEE), a rare virus infection that usually strikes only four or five Americans a year. EEE is a highly lethal disease; up to half its victims die, and many survivors are left with long-term brain damage. And although the child lived, he will never be the same again. As one Harvard scientist puts it, "This is a ruined life."

The boy caught EEE not from horses, as its name implies, but from something that every playful child encounters over and over again in the course of a summer: a mosquito bite. (Two other Massachusetts residents, a seventy-five-year-old man and a sixty-nine-year-old woman, also contracted EEE in 1990; the man died.) EEE is one of nearly one hundred human diseases that fall into the category of arboviruses, a shorthand for *ar*thropod-*bor*ne viruses. (Arthropods are invertebrates with jointed legs and segmented bodies; the more colloquial term, insects, refers only to those arthropods—like mosquitoes—with wings, bodies in three segments, and three pairs of legs. Other arthropods are crustaceans, such as shrimp, crabs, and lobsters; arachnids, such

**FIGURE 5. Life cycles of arboviruses**

| DISEASE | SYMPTOMS | TRANSMISSION CYCLE | VECTORS |
|---|---|---|---|
| DENGUE FEVER | High fever, bone and joint pain (breakbone fever), intense headache, skin rash, small hemorrhages just under the skin, nausea, vomiting, swollen glands, fatigue, depression; in dengue hemorrhagic fever, internal bleeding may lead to shock | 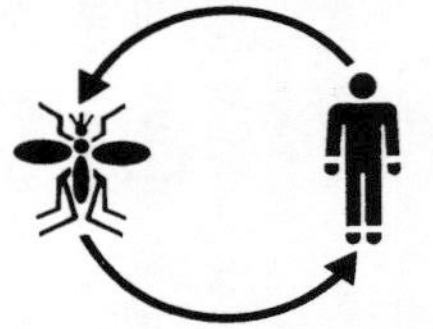 | AFRICA AND AMERICAS: *Aedes aegypti*; ASIA: *A. aegypti, A. polynesiensis, A. albopictus* |
| YELLOW FEVER | Sudden-onset fever, headache, backache, nausea, vomiting, slowed pulse, reduced urine production, low white-cell count; later, bleeding from nose and mouth, vomiting blood (black vomit), jaundice, lesions of the liver, kidneys, and gastrointestinal tract | 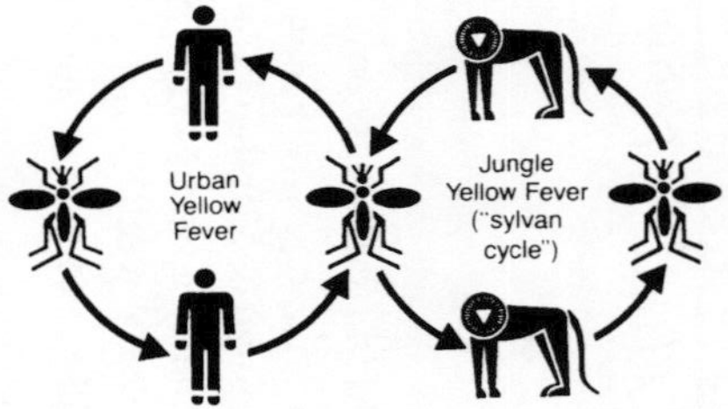 Jungle Yellow Fever ("sylvan cycle"); Urban Yellow Fever | AFRICA: *Aedes africanus, A. aegypti, A. bromeliae,* and others; AMERICAS: *Haemagogus* species |
| EASTERN EQUINE ENCEPHALITIS | Headache, drowsiness, fever, vomiting, and stiff neck, rapidly progressing to tremors, mental confusion, convulsions, and coma | 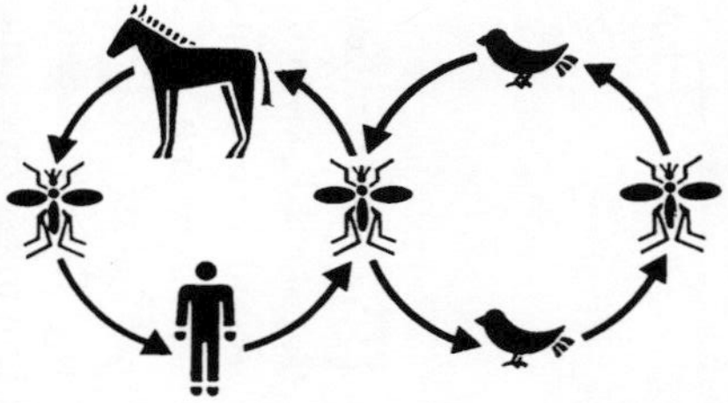 | EASTERN U.S.: *Culisera melanura* maintenance vector; *Aedes vexans* and *Coquillettidia perturbans* bridge vectors |

as spiders, scorpions, ticks, and mites; and myriapods, like millipedes and centipedes.) Among the arboviruses are some of the scariest viruses known; they are transmitted by vectors that are ubiquitous and often unavoidable, and they commonly lead to devastating illnesses. For people in certain regions of the world, arboviruses are as inescapable as the sunrise. Their special terror is that they are transmitted by bites from insects or other animals so small they sometimes cannot be seen, so pervasive they sometimes cannot be avoided.

Some arboviral diseases are given bland names that flatly describe their symptoms: yellow fever leaves its victims jaundiced, or "yellow," because of liver failure; tick-borne encephalitis is a brain (*encephal-*) inflammation (*-itis*) that is carried by a tiny tick. Yet these names manage to convey, even in their dreadful ordinariness, a horrid image of death. Then there are the arboviruses that sound every bit as frightening as they are. Because so many were discovered in the tropical regions of the world, even names that merely identify the towns where they were first seen sound distinctly exotic to Western ears: Oropouche and Rocio viruses from Latin America; Kyasanur Forest disease virus from India; and from Africa, Obodhiang virus, Orungu virus, Chikungunya (named for the Tanzanian word for "that which bends up," as victims' limbs do because they are in such terrific joint pain), and O'nyong-nyong, from the East African tribal word for "joint breaker."

But scary as they are, arboviruses generally enjoy no advantage in invading human beings. As with so many other viruses, when an arbovirus gets into a person—and causes encephalitis or hemorrhaging or some other horrid condition—it often happens only because that person somehow accidentally bumped up against a virus that was sticking to its own life course. The added twist, for the arboviruses, is that their life cycle usually *does* depend on so many other species. It's just that human beings, and indeed most large mammals, are not typically part of the scheme. (There are a few notable exceptions to this generalization: yellow fever and dengue fever do depend on humans for perpetuation.) What arboviruses typically need are two things only: an arthropod vector and an avian or mammalian reservoir.

The vectors and their reservoirs make arboviruses especially portable—which, in turn, means that many of the emerging viruses under study today are arboviruses. If an arbovirus requires only a vector population and an available, nonimmune reservoir species, it can move from one hemisphere to another as easily as ships can cross the oceans.

Indeed, such flexibility accounts for the introduction of many new arboviral diseases even in our own lifetime, and the certainty that some scientists feel that the next emerging virus, at least in the United States, will be a new (to the United States) arbovirus carried by a particularly pesky little bug.

Mosquitoes are the arthropods most often involved in human arboviral diseases, but the viral vector may also be a biting midge, a tick, or a sand fly. The reservoir can be any animal—rodent, bird, ruminant, primate—in which the virus is able to multiply many times over, usually in the animal's lymphoid or connective tissue. Inside the reservoir, the virus proliferates, or "amplifies," so that eventually there is so much active virus in the bloodstream that any arthropod that bites it is bound to get a good healthy dose.

In arboviruses, as with perhaps no other group of viruses, how we live affects what we catch. Colorado tick fever virus, for instance, is most likely to infect adult males, since the prime time for catching it is while hunting or fishing in the forested regions of the Rocky Mountains. LaCrosse encephalitis is more likely to infect young boys, since the mosquito that carries the virus tends to breed in water that accumulates in the joints of trees—especially in good tree-climbing trees. As recreational patterns change, the pattern of arboviral disease transmission might change too. Who knows whether more young girls will start getting infected with LaCrosse virus as equal opportunity grants them access to treehouses that were previously posted, No Girls Allowed.

Social changes might have other implications for the spread of these diseases. Arboviruses are most widely disseminated in the world's tropics not simply because that's where the mosquitoes live, but also because that's where people are exposed to mosquitoes day and night. On steamy hot nights, many people spend hours trying to cool off outdoors—where they are prey to the bites of avid nocturnal feeders such as the *Culex* mosquito, the vector for Japanese encephalitis in Asia and St. Louis encephalitis in the United States. In the Western world, people tend to avoid nocturnal mosquitoes behind screened windows, air conditioning, and a life-style that keeps them inside on summer nights watching television. But this is likely to change as homelessness becomes an increasing problem. Already, a mini-outbreak in southern California of malaria—which is caused by a protozoan, not a virus, carried by the *Anopheles* mosquito—has been traced to migrant workers forced to live in unprotected shacks, in conditions so squalid they resemble the worst urban slums of the developing world.

Because they coexist in such superb syncopation with people's habits, the populations of arboviral vectors—and, therefore, the incidence of whatever disease they carry—rise and fall with the slightest change in human behavior. A few mosquito vectors breed around households that unwittingly provide the fodder for their larvae; if a Coke can is left in the backyard too long, or if cracks along the mortar are allowed to spread, one of these domesticated mosquitoes may soon start breeding in the tiny slurps of water that collect in the gaps. Once that happens, the pests will persist. These arthropods tend to be highly domesticated, actually preferring to live in and near houses, rather than living in the wild—a good strategy, after all, since their main sources of protein, human beings, live inside those houses. So arboviruses are at the top of anyone's list of emerging viruses, teetering as they do on the edge between the natural and the domestic world.

With their vectors so plentiful, and being so near the animal hosts that allow them a chance to amplify, arboviruses can afford to be choosy about the species with which they will coexist. And choosy they are: most can be carried in only one or two kinds of arthropods, unable to exist even in a very closely related species. Chikungunya virus, for instance, can replicate in an *Aedes aegypti* mosquito, but not in *Culex tritaeniorhynchus*. The reason for this specificity is that the virus has evolved in careful synchrony with both reservoir and vector. It must be capable of invading the arthropod's gut, where it can continue to amplify, and then must be able to move to the salivary glands, where it waits until the arthropod bites another victim. If the gut or the salivary glands do not accept the arbovirus, the transmission cycle ends then and there. Similarly, arboviruses tend to be quite restricted in the range of reservoir animals they can infect. EEE virus grows in most species of migratory birds, but a related virus, LaCrosse encephalitis virus, grows only in small mammals. Either one causes disease in cattle or people, but never giraffes.

The mosquito is an especially good viral vector because of its odd biting ritual. Like a major league ballplayer, a mosquito always spits a little before it gets down to work; it wants to prime the biting target with its own saliva, which contains an antiplatelet component that keeps the victim's blood flowing freely into the mosquito's proboscis. The mosquito takes in an enormous amount of blood in a single feed, ballooning to four times its normal weight. It needs about two days to digest the gargantuan meal, during which time the mosquito uses the protein from the blood to make eggs. (Only female mosquitoes bite.) Then it lays its eggs and is ready for more blood. The mosquito is like a flying hypodermic needle, capable of transporting the virus from one spot to another, the way bees carry pollen among flowers. But it is also something more, since it plays not only a physical role in the virus's life cycle but a biological role as well. By amplifying the virus in its salivary glands, the mosquito increases the odds that the virus will be passed on to a new host. Amplification increases the likelihood that the mosquito's bolus of saliva, delivered as a warm-up during the course of a later blood meal, will indeed carry enough virus to get transmitted.

It seems a rather tenuous system for survival. Even when conditions for transmission are perfect, as in the virology laboratory, the majority of mosquitoes exposed to the virus will not ultimately transmit it. Suppose you begin with a laboratory monkey that is capable of amplifying the virus that causes dengue fever. Then you put the animal in a cage with lots of *Aedes aegypti* mosquitoes, a species that likes to bite monkeys (mosquitoes can be as particular as viruses are) and is capable of amplifying dengue fever virus. If you get those mosquitoes out of the cage after a week or so, you can inoculate their tissues into a culture system and actually measure what percentage of them harbors dengue fever virus. And if you examine their salivary glands, you can see what proportion of infected mosquitoes is capable of passing on this virus next time they bite another host. Most experiments conducted in this way have found that about 80 percent of the mosquitoes exposed to infected monkeys become infected themselves; of those, about one-third to one-half have virus disseminated to their salivary glands, meaning they are capable of transmitting infection to another animal. So for every hundred mosquitoes exposed to a dengue-infected monkey, no more than forty—and usually far fewer—end up with the virus poised for delivery in the salivary glands.

In the real world, the odds are even more stacked against the virus. "In a field situation when you look for infected insects, if you find one in two thousand or one in four thousand infected, that's a high rate," says Robert Tesh, a virologist and entomologist at the Yale Arbovirus Research Unit (YARU). The rate increases significantly during large epidemics, to maybe one in two hundred or three hundred, but still the vast majority of mosquitoes carries no virus. In addition, the arbovirus must race against the foreshortened time of the mosquito's life span. Most mosquitoes live no more than half a week, yet arboviruses require at least two or three days to be amplified in the mosquito's gut. The only hope a virus has of being transmitted—since the mosquito must remain spry enough to bite a second time after amplification in order to pass on the virus to a new host—is to have had the blind good fortune of replicating inside an incredibly long-lived arthropod.

How can arboviruses withstand such overwhelming odds? Why is it that, despite these constraints, they have thrived for many centuries, dating back at least as far as the Han dynasty in China, two hundred years before the birth of Christ? The most likely explanation is that mosquitoes bite with such frequency, such ferocity, that the number of total possible transmissions is huge. It may be that only one bite in two thousand carries a virus, but with many millions of bites involved, that provides for plenty of loaded bites. Another reason for the arbovirus's adaptive success has been its mobility. Arboviruses that amplify in birds or bats are especially able to expand their range far beyond the range of non-arboviruses, which require the intimate association of infected and uninfected host. Most viral outbreaks occur only in geographically restricted animal herds or human communities; arboviruses can spread across whole continents. Unlike horses or people, birds cover enormous amounts of territory in their lifetimes; vector-borne infections that amplify in birds will thus be more widely disseminated than infections that are transmitted only by direct contact. Think of malaria, which is not a viral disease but is transmitted by mosquitoes. It infects an estimated three hundred million people a year, killing about a million and a half of them—making it the single most pervasive infectious disease on earth. Only a vector-borne disease could accumulate such a notorious track record.

The virus that causes EEE has a life cycle that is typical of many arboviruses. It involves a single species of mosquito as vector, a few types of wild birds as reservoir, and an occasional leap, via a different mosquito species, into "dead-end hosts" like horses and human beings. *Culiseta melanura* is the maintenance vector of EEE; it is in this mos-

quito that the virus travels to the salivary glands and continues to proliferate. The reservoir animals are migrating songbirds, such as the
red-wing blackbird and the snowy egret, in which the virus can persist
and amplify while causing the birds no harm. (This lack of harm to
the reservoir host is crucial. A dead or dying animal is usually not out
in the open and available for biting. It thus becomes a different kind
of dead-end host.) The cycle works this way: A mosquito bites an infected bird. The virus passes from bird to mosquito and multiplies in
the mosquito's gut. Then the virus travels to the mosquito's salivary
glands and passes into a new bird when the mosquito bites it, and the
cycle repeats indefinitely. All that is needed is a good supply of nonimmune birds (the virus simply dies in birds that are immune through
prior exposure), a bit of marshland, and enough spots in the marsh for
the *Culiseta* to go through its own quirky life cycle, laying its eggs in
the only spot it likes: under the roots of tall swamp trees that have been
undercut or partially overturned by high winds.

    *Culiseta* is not only choosy about where it breeds; it's also quite
particular about what it bites. Its narrow biting range, restricted almost entirely to birds, makes it an especially efficient vector for EEE.
"The vector has to bite the reservoir host two times," says Andrew
Spielman, a vector biologist at the Harvard School of Public Health.
"So if either bite is wasted on a non-reservoir host, the other bite is
also canceled out." The finicky tastes of *Culiseta* thus at least doubles
the efficiency with which it can transmit the EEE virus. Very few bites
are ever wasted; the cycle just continues as bird to mosquito to bird
to mosquito to bird. Occasionally, though, this cycle is disrupted, and
EEE begins to infect other animals—and actually make them quite
sick.

    This break in efficiency is not the doing of the little *Culiseta*,
though. It's the work of two quite different mosquito species, the type
that will bite anything that moves: *Coquillettidia perturbans* and *Aedes
vexans*. Their very names are testimony to their "perturbing" and "vexing" biting habits. In the life cycle of EEE, these two mosquitoes serve
as "bridge vectors"; they span the gap between birds and mammals.
*Coquillettidia perturbans* breeds at the base of marshland cattail plants—
almost as specialized a locale as the upturned tree roots required by
*Culiseta*. *Aedes vexans* breeds in small puddles that form after unusually
heavy summer rains. In years in which either cattails or puddles occur
in higher-than-normal amounts, scientists have learned to anticipate

an increase in the population of these bridge vectors, and therefore to be on the lookout for EEE among local horses, flocks of domestic pheasants, and, ultimately, human beings.

Such years are not only bad news for people; they're bad news for the virus too. A mammalian host is a dead end for the EEE virus: the cycle of transmission is stopped because horses and people cannot amplify it, so they will never become sufficiently viremic (having virus in the bloodstream) to pass on an infection through a second mosquito bite. Perhaps that is why the equine and human versions of this encephalitis are so destructive. Evolutionary forces never had a chance to mitigate its ravaging effect on people.

Because of the vagaries of their transmission cycles, arboviruses are especially susceptible to external control. Eliminating the vector itself is one method. This can involve techniques as primitive, or as elaborate, as the cultures in which the diseases are occurring. In the tropics, it can be something no more elaborate than getting patients indoors. In many Latin American cultures, for instance, sick people rarely take to their beds; even at the height of their illnesses, they often sit outdoors and intermingle with others. But having an acutely ill viremic person outdoors while the mosquitoes are biting only perpetuates the cycle of disease transmission. Sometimes all that would be needed to help control the spread of disease would be to send all victims to bed and make sure there is intact mosquito netting around them. If no patients were available to bite, new mosquitoes would not take in the virus in the first place.

Other life-style changes also help limit the spread of disease. The most important changes involve open water containers, where many disease vectors like to lay their eggs. If no standing water exists, mosquitoes cannot breed; that is why the incidence of many arboviruses is so closely related to a season's rainfall. In certain regions of the world, tiny pools of water are inevitably created as part of the daily routine. Water barrels sit uncovered just outside the kitchen door; small vases of flowers lie on gravestones in rural cemeteries; used automobile tires and old tin cans strewn in urban backyards collect water with each new storm. It is possible, just by covering up such sources of water, to interrupt the breeding capabilities of many mosquitoes, especially *Aedes aegypti*, which lays its eggs exclusively in artificial containers rather than in natural puddles or water holes. An ancient folk ritual of the Tamils, a language group from southern India, is silent testimony to the wisdom of this simple prohibition. The Tamil custom is to bring water into the house only once a day, and not to store water at any other time. When Tamil laborers were brought into Malaya to work on plantations during the 1950s, dengue fever and malaria were endemic in the region. But the prohibition on storing water kept the disease rate among the Tamils far lower than that of their Malay or Chinese co-workers, all of whom lived under similar conditions but who stored water indoors.

After all the patients are isolated and all the breeding grounds are removed, arboviral disease control gets more aggressive. This includes, as a first line of attack, the spraying of insecticides to kill off mosquito vectors. The nice thing about spraying is that it seems to have such a huge and immediate impact. In 1901, for instance, Cuba was in the grip of a yellow fever epidemic that no one could contain. In swept the United States Army Medical Corps, led by physicians Walter Reed and William Gorgas, and every spot in the city of Havana was fumigated. Within nine months, the mosquitoes were gone, and the plague was over. (Spielman of Harvard thinks the real hero in that story was Teddy Roosevelt, who stopped the steady flow of Spanish immigrants into Cuba in 1898, with the Spanish-American War. Without immunologically naive humans to infect, says Spielman—in an argument many of his colleagues reject—yellow fever would have petered out anyway even if the *Aedes aegypti* population had been entirely untouched.)

But though the results seem dramatic, they are usually only temporary. Indeed, the spraying itself might exacerbate the problem. Since no spray eliminates every single mosquito, those small numbers that do survive proliferate rapidly; they have no competition for food and breeding spots in subsequent generations. Many of the mosquito's natural predators might also be killed off during spraying, further reducing the ecological pressures to keep mosquito numbers down. And the mosquitoes that live through massive spraying programs with organophosphate pesticides such as DDT probably had some inborn, genetic resistance that accounts for their surival. They become the only strain remaining, and when they breed they pass on this insecticide resistance to their offspring. In the case of *Aedes aegypti*, there's another complication: the mosquito hides in such remote corners that it is often out of the reach of aerial sprays. "This particular mosquito lives indoors, in dark corners that are relatively immune to being reached by the insecticides," says Thomas Monath, a virologist now at Ora Vax, a biotechnology firm in Cambridge, Massachusetts. *Aedes aegypti* tends to live in closets or other enclosed spaces, he says, and most mosquito spraying barely filters to the rooms of a house.

In the long run, then, insecticides often lead to an increase, rather than a decrease, in mosquito populations, and the majority of those new mosquitoes are resistant to whatever sprays human beings have at their disposal. This paradox of pesticides has been proved over and over again in this century, throughout Central and South America. In the years since the Havana sweep, massive spraying was used to deal with plagues of dengue fever and yellow fever—both of which are carried by the *Aedes aegypti* mosquito—in Panama, Brazil, Ecuador, Peru, Cuba again, and Mexico. And though it seemed for a time as though *Aedes aegypti* had been vanquished from the continent, beginning in the early 1980s, the vector began a rapid and steady population increase that continues to this day.

In light of the failure of mass spraying programs, some investigators are now advocating a more precise approach to mosquito warfare: instead of pulverizing the vector, just try to defang it. The first attempts along these lines involved releasing sterile males, whose sperm-producing capability has been destroyed by huge doses of radiation, into a targeted region. This was the strategy used in the 1970s to stem the influx of Mediterranean fruit flies into California. Sterile male "decoys" were expected to deter females from mating with flies that would actually fertilize their eggs, thus subverting a whole generation. But it didn't work. The laboratory-reared flies did not know how to fight for the females, so they rarely mated successfully; the radiation they had been given was in such high doses that the flies could not fly; and the female flies responded to the subterfuge by laying an inordinate amount of eggs.

Now scientists have gone reproductive tinkering one better. They are trying to manipulate the genes of important arbovirus vectors in such a way that they become "vector incompetent." Once released into the environment, the hope is, these miniature mutants will pass on their vector-incompetent genes to subsequent, similarly incompetent, generations. Most of the genetic research so far has been done on mosquitoes, primarily because they are the most common vector for arboviral, malarial, and other human diseases. Mosquitoes have only three pairs of chromosomes, and knowing which genes sit at which sites on those chromosomes—what the scientists call gene mapping—should enable molecular biologists to introduce specific mutations at just the right spot.

Genes control two aspects of vector competence: the mosquito's ability to amplify a virus in its gut and its ability to move the virus to its salivary glands. The investigator's first step, then, is to find which genes control these functions, by comparing mosquitoes that are vector competent with those that are not, and identifying genetic differences. Two closely related species of *Aedes* mosquito, for example, are capable of carrying the LaCrosse virus, which causes a serious form of encephalitis in thousands of American children a year, mostly in the north-central United States. But only one species, *Aedes triseriatus*, is capable of transmitting LaCrosse virus to humans. Its cousin, *Aedes hendersoni*, can take in the virus and even amplify it, but the virus never gets into its salivary glands, so the mosquito is unable to transmit the virus to humans when it spits on them before a bite. The reason *Aedes hendersoni* is harmless, it turns out, is that the pores in its salivary glands are so small that they screen out viruses. Scientists at the University of Notre Dame have managed to identify the one or two genes that code for small pores in this species. Next, they plan to transmit the small-pore genes into *Aedes triseriatus*, either through selective breeding—the traditional form of genetic engineering—or by literally moving the relevant genes into fertilized eggs. The idea is to change the gene pool of the vector-competent *Aedes triseriatus*, giving subsequent generations such small pores that the entire species becomes vector-incompetent. The ultimate test of whether it works, of course, will be to release the genetic hybrids into the wild, to see whether they reproduce successfully enough to replace the native strain.

Other scientists are trying to map the genes for other mechanisms of vector competence. One relevant mechanism is the mosquito's own immunelike response to a virus. Because mosquitoes can develop a kind of immunity to viruses to which they have been exposed, they usually can amplify a virus only once; after that first infection and transmission, they manufacture proteins in their gut that inhibit the replication of any virus. This restriction usually has no real impact on the virus's life cycle, since most mosquitoes live for less than a week and most viruses take several days to amplify. As far as the virus is concerned, once is enough; indeed, once is about all it ever gets. But what scientists would like to do is to stimulate this quasi-immune response in mosquitoes *before* they have been exposed to any virus and *before* they have had the days they need to amplify that virus and pass it on to host number two. At Colorado State University, entomologists are trying to isolate the gene for this gut immunity, and they then hope to splice the right genes into vector-competent mosquitoes and, as with the Notre Dame research, release the mutants into the wild.

That release into the wild is the great unknown in this type of experimentation. So far, mosquitoes that are genetically engineered have proved to be extraordinarily fragile, and they don't survive for long outside the protected environs of the entomology lab. That, after all, was the fatal flaw in the medfly experiment, when thousands of sterile males were released in the 1970s to combat a plague of crop-damaging fruit flies in California. It's one thing to breed laboratory mosquitoes that are missing a particular gene. It's quite another to release them into a natural habitat and hope they will not only survive but also eventually edge out the species already there. "The whole history of entomology is of people trying to develop something in the laboratory and then releasing it in nature; the laboratory insects just don't survive," says Robert Tesh of Yale. "This happens with animals too. If you took a white laboratory rat and threw it out in the ghetto where rats live, it would disappear among the rats that have been out there every day, scrounging and fighting."

But George Craig, an entomologist at Notre Dame, says laboratory mutants need not subjugate themselves to the native strain; indeed, in order eventually to outnumber the non-mutants, all they need is that their mutation be "neutral" in terms of their ability to survive in the local habitat. By this he means that the genetic alteration confers neither an advantage nor a disadvantage in terms of survival and adaptation. Craig says he proved this during the seven years he worked in Mombasa, Kenya, in the 1970s, when he switched the eye color gene of *Aedes aegypti* mosquitoes and released them into the wild. After just one breeding season, he says, "all the mosquitoes there had red eyes." Perhaps a vector-incompetence gene like small pores would prove to be as neutral ecologically as the red-eye gene. The only way to know for sure is to try it. Experiments of this sort are expensive, though, and Craig says he has been having trouble getting government grants to conduct research that does not have an obvious and immediate public health payoff.

Arboviral diseases seem to materialize virtually without warning. Sometimes it takes years of dogged detective work to figure out how they emerged, and why they emerged at a particular time and place. With the background of experience, scientists can anticipate what new diseases are waiting in the wings—but even when they know what's coming next, they are often paralyzed, through lack of money or lack of collective will, to do anything to prevent it.

One of the most perplexing new diseases to emerge in the western hemisphere seems, in retrospect, to have been quite easy to predict. The punch line is that a change in the major cash crops in Brazil led to the emergence of a new, fast-spreading arbovirus, and that its dissemination was amplified by new roads being built into the Amazonian rain forest. But the facts were not all pieced together until the 1980s, nearly thirty years after the mystery first arose.

It began with a dead sloth on the side of the road. The year was 1960, a major highway was being cut through the rain forest linking the city of Belém and the new capital of Brasília, and the young American physician working at the Belém virology laboratory was Robert Shope, son of the esteemed virologist Richard Shope of The Rockefeller University. In the best of times, the Belém–Brasília highway was a mass of dust and stone, and in the worst of times a muddy morass. At any time, that sloth did not belong there. Shope and his colleagues performed an autopsy on the sloth, which they decided had died of illness rather than accident because of where it was found, and in its blood was a rare virus known as Oropouche. This virus was named after the village in Trinidad where it had been identified six years earlier. Oropouche virus had never been associated with any large outbreak of disease, animal or human, so the scientists mentally filed away the sloth's infection as a curiosity, to add to their inventory of rain forest viruses that were being inadvertently uncovered by construction through the jungle.

But just one year later, this same unusual virus began turning up in people. In 1961, a plague of flulike illness swept through Belém; more than eleven thousand people came down with high fevers, debilitating headaches, and severe muscle aches and pains. Almost everyone eventually recovered, but the scientists working at the Belém Virus Laboratory wanted to find out what caused it. As part of a network of investigators funded worldwide during the 1950s and 1960s by The Rockefeller Foundation, scientists in Belém considered it their responsibility to isolate, identify, and classify as many previously unknown viruses as they could—an enterprise derided by some as "stamp collecting," but one that has undoubtedly advanced international awareness of a good many viruses of public health significance.

The scientists began searching out blood samples from hundreds of recovered patients. To do this, they enlisted the help of field epidemiologists. "Field epidemiologist is a euphemism," wrote two noted virologists who held that title during the study of Oropouche, "for the person who wades through the mud, questions the suspicious local inhabitants, finally locates the patient again in a little shack deep in the forest and persuades him, now completely recovered from his illness and hardly remembering that he had been sick three weeks earlier, to part with another specimen of blood." In several of these hard-won specimens, the scientists found the Oropouche virus.

Under an electron microscope, the virus appeared to be arthropod-borne; it structurally resembled other known arboviruses. So the scientists began surveying the local mosquito population for signs of Oropouche virus in *their* blood. This involved catching mosquitoes in the field. And if "field epidemiologist" is a euphemism, "mosquito collector" is a downright confabulation, since that person is nothing more than human bait. Mosquito collectors were young Brazilians sent out into the rain forest, naked to the waist, who stood there while the mosquitoes came to feed. "We taught the men how to catch the mosquitoes in little vials, hopefully before they were actually bitten," says Robert Shope, who is now director of the Yale Arbovirus Research Unit, the largest nongovernment U.S. lab devoted to the study of arboviruses. But bitten they were. Bites did not usually pose a problem for natives, who had been exposed their whole lives to these mosquitoes and the viruses they carried. But sometimes surprises erupted. One young Brazilian working for the scientific team volunteered to go into the rain forest after dusk to catch *Culex* mosquitoes, a nocturnal species. But he had previously been in the forest only in the daytime. "He was sick within a week," Shope says. "What we had suspected, and found out to be true, was that the different species of mosquitoes carried different viruses." Unfortunately, the young man had never developed immunity to those different viruses.

After the mosquitoes were collected, scientists in the Belém laboratory ground them up with mortar and pestle, mixed the remains with a sterile solution, and injected the liquid into the brains of laboratory mice. If the mice sickened or died, their brains were examined under the microscope to see if any virus could be isolated. The results were confusing: mosquitoes trapped in the rain forest did indeed carry a virus that killed the mice, but those caught in town, where the disease was spreading rapidly, did not. Some other insect arthropod must be responsible for the urban transmission of Oropouche.

By the time Shope headed back home to the United States in 1967—his wife, he says, had had enough of the hardships of trying to raise three babies in Brazil—he still did not know what the Oropouche vector was. The mystery wasn't solved until 1980, when a rather bizarre set of circumstances brought a new epidemic of Oropouche fever to northern Brazil. Once again, most of the mosquitoes tested from the immediate region were free of the virus. So scientists began testing other arthropods that are known to carry viruses, such as ticks and mites. They finally found Oropouche virus in a minuscule arthropod known as a biting midge (*Culicoides paraensis*, which is called a sand fly in the United States). And what had changed in 1980 to account for the sudden outbreak? That took some more detective work. The epidemic eventually was traced, albeit indirectly, to the industrialized world's adoration of chocolate. As cacao, whose seeds are used to make chocolate, became a more important cash crop in the impoverished regions of northern Brazil, more and more land was turned over to cacao trees. When the cacao hulls were discarded, they piled up into huge mountains, each hull containing a mini-pool of collected rainwater. Those pools turned out to be the perfect breeding ground for the biting midge. As the piles of hulls grew higher, the midge population grew too. And as the midge population increased, so did an individual human's odds of being bitten by a midge that carried the virus for Oropouche.

Scientists tracking emerging viruses must look at many factors in trying to predict what is coming next: they must consider not only scientific and health factors but shifts in the political, economic, and social climates as well. If the emerging virus is also an arbovirus, there's that much more to consider; forecasting involves an understanding of how humans live and what insects or animals they encounter, plus an understanding of the life cycles of the vectors and the reservoirs themselves. In North and South America, all these factors taken together indicate that conditions are ripe for the emergence of a particularly worrisome arboviral disease: dengue hemorrhagic fever. One potential mosquito vector has recently invaded the United States and Brazil, and it may be only a matter of time before it becomes infected with the dengue virus and passes it on to its first native victim.

Dengue hemorrhagic fever is the severe form of a condition known as dengue fever, a nonfatal infection that occurs in cycles in Southeast Asia and South America. The name might have been coined in Havana in 1828, from the Spanish words *el dengue* meaning "affectation," probably because the joint pain it causes makes victims assume foppish poses. Or it might derive from the Swahili *dinga* or *dyenga*, which means a sudden cramplike seizure—a term used to describe epidemics noted in Zanzibar in 1823 and 1870. Dengue fever has been observed in North America too, as long ago as the summer of 1780, when the prominent American physician Benjamin Rush, one of the signers of the Declaration of Independence, wrote a detailed clinical description of an outbreak of dengue fever in Philadelphia. In those days, sailing ships frequently traveled between the Caribbean, where dengue was endemic, and the colonial port cities of Baltimore, Boston, Charleston, Philadelphia, and New York. Not much was required for outbreaks to occur; just a few sailors infected with dengue, and the release of hundreds of *Aedes aegypti* mosquitoes from the big open barrels of water that were stored on board for the journey. The mosquitoes died out every winter, since they could not tolerate the cold, but their reintroduction again the following spring with a new fleet of ships ensured that dengue fever would remain a problem for years.

Once public health experts became aware of the pattern, simply covering the ships' water barrels was enough to limit the annual influx of *Aedes aegypti* into the United States. Dengue fever went into a sharp decline in North America, except for the occasional case of "imported dengue" brought in by someone who had traveled to an island in Oceania or the Carribean and returned home infected, usually even before symptoms began. Meanwhile, in Asia, Africa, and Latin America, endemic dengue fever continued to spread. Every year, an estimated 100 million people around the world are infected. They suffer from severe headache, eye pain, and muscle and joint pain so disabling that dengue is known in Africa by the nickname "breakbone fever." Recovery is usually complete within about ten days. But what worries public health officials now is that a prior infection with one strain of dengue fever virus seems to predispose people—especially children— to developing a much more dangerous form of dengue fever after ex- posure to a different viral strain. This special manifestation of dengue, known as dengue hemorrhagic fever, first emerged in the 1950s. It leads to high fever, internal bleeding, leaking of the capillaries, coma, and shock. About 10 to 15 percent of affected people die. And it is the reason why concern about the dengue fever vector is steadily growing around the world.

The emergence of dengue hemorrhagic fever can be traced to the worldwide dissemination of four distinct varieties, or serotypes, of the dengue virus. Of the four dengue serotypes—known prosaically as types 1, 2, 3, and 4—two or more began to circulate in the same place at the same time, beginning in the mid-1950s. The coexistence of more than one serotype in the same geographic region is a necessary con- dition for the emergence of dengue hemorrhagic fever, which tends to be restricted to people encountering a second dengue infection when the serotype involved in the second infection is generally different from the one that caused the first. This condition of sequential infec- tion with different serotypes is, of course, far more likely to be the case if many different serotypes are making the rounds.

If you laid the RNA of each dengue serotype end to end and counted down the nucleotides, there would be only about 50 percent similarity between any two types—about the same degree of homol- ogy that exists between one dengue virus serotype and the related virus that causes yellow fever. (One wonders why the types, if they are so dissimilar, are classified as the same virus at all; the short answer is that scientific convention is difficult to change.) The close-but- imperfect homology among different serotypes of dengue virus yields to a close-but-imperfect fit between one dengue *antibody* type and an- other. This throws off the immune response to a second attack of dengue by a different viral serotype, which the established antibodies cannot quite neutralize. As virologists see it, dengue hemorrhagic fever is a classic case of the body's immune defenses doing more harm than good. "It's at least a hundred times more likely that you're going to get the severe form of dengue fever if you've previously been infected," says Tom Monath. The odds of getting dengue hemorrhagic fever from a first infection with dengue virus are about one in twenty thousand, he says; after a person has been "sensitized" by a prior dengue infection, the odds shoot up to about one in one hundred to three hundred.

Suppose a young girl in Havana is infected with dengue type 1. If she recovers, she will continue to have antibodies to dengue type 1 circulating in her bloodstream. If, three years later, she is exposed to dengue type 2, her type 1 antibodies will offer her no good protection. It is almost as though she is encountering a totally new virus—almost, but not quite. There is still that 50 percent homology, still that tendency for the type 1 antibody to cross-react with pieces of the type 2 virus. Instead of a complete embrace that would sweep the virus off its feet and totally immobilize it, the antibody manages only a clumsy sidesaddle hug, clinging onto the virus like two people tied up in a potato sack for a three-legged race. The result is that the antibody binds with the virus but is unable to neutralize it. It forms instead a chimerical molecule—half antibody, half infectious virus—known as an immune complex. Certain cells in the immune system, the macrophages, recognize the antibody half of this odd hybrid. They welcome the immune complex as though they were getting ready for the next step in the defense strategy, the step of engulfing and destroying any molecule that the antibody presents them with. But they do not recognize that the unbound half of the immune complex is free-floating virus, capable of infecting and destroying the macrophage. So the macrophages, instead of destroying the immune complex, are themselves destroyed. The result is the release of chemicals that can devastate the immune system and the circulatory system, causing the symptoms of dengue hemorrhagic fever. If this little girl does not make it to a hospital in time, she runs a one in ten risk of dying.

In 1981, dengue hemorrhagic fever—by then the leading cause of pediatric hospitalization in Southeast Asia—struck for the first time in the western hemisphere. An outbreak in Cuba infected 344,000 people and resulted in 116,000 hospitalizations during one terrible three-month period. Subsequent outbreaks in Latin America were reported more recently, in Ecuador (1988), Venezuela (1989), and Peru (1990).

Not everyone believes that dengue hemorrhagic fever requires the circulation of two dengue serotypes to become a problem. Leon Rosen

of the University of Hawaii cites instances of dengue hemorrhagic fever that occur even on first exposure to a dengue virus. According to
Rosen, an outbreak of dengue type 2 on the isolated Pacific island of
Niue in the early 1970s was associated with some cases of dengue hemorrhagic fever, even though the island had experienced no dengue outbreaks for at least twenty-five years. Since the condition existed in
people under the age of twenty-five, Rosen wrote, "it could not have
been the result of sequential dengue infections." Similar instances of
dengue hemorrhagic fever in people who had never before been exposed to dengue make Rosen question the theory that the hemorrhagic
syndrome could be accounted for by a lopsided immune system
reactivation.

If closely related antibodies do in fact turn dengue fever into dengue
hemorrhagic fever, some scientists question the wisdom of using the
vaccine against another closely related virus, yellow fever, in regions
where dengue fever is endemic. Now that a few cases of urban yellow
fever have been observed in Brazil, that country's health ministry is
considering a massive inoculation of more than 100 million citizens.
The critical question, though, is whether the yellow fever antibodies
that would be generated would behave, in Brazilians' bodies, like antibodies to a sort of fifth dengue serotype. In other words, would vaccinated people exposed to a first infection with dengue fever be at high
risk for hemorrhagic complications? "What's going on now in Brazil is
a true experiment of nature," says Monath. "We just don't know what
would happen to urban populations protected against yellow fever
with a vaccine and then exposed to dengue infection."

In the United States, the dengue fever issue is a little more straightforward. The reason American officials anticipate dengue fever as the
next new viral disease is that a new, especially ferocious vector has
already arrived. *Aedes aegypti*, which carries dengue fever virus in
South America, has for centuries lived in the warmer states of the
United States. But now a new, temperate-climate mosquito is replacing *Aedes aegypti*, and it is reputed to have a more ravenous biting habit
and a greater ability to withstand American winters. The new mosquito, the Asian tiger mosquito, first sailed into Texas in that shipment
of used Japanese automobile tires, and its range has expanded at least
as far north as Illinois, with no signs of slowing.

Laboratory work with the tiger mosquito, or *Aedes albopictus*, shows
that it is capable of amplifying and carrying a great number of viruses

in its gut and salivary glands: not only dengue and yellow fever, which were previously known about, but also some particularly American viruses such as LaCrosse encephalitis and a new virus called Potosi virus (named after the Missouri town in which it was isolated), whose clinical significance has yet to be discovered. In all, *Aedes albopictus* is thought to be capable of carrying at least fifteen viruses. "In my opinion, LaCrosse could be much worse than dengue," says George Craig, who is trying to get the government worked up about the disease threat of tiger mosquitoes. "It's already in place. It just needs a new vector." The traditional vector for LaCrosse encephalitis, which causes about one hundred cases of obvious infection annually and an unknown number of subclinical infections with subtle neurological effects, is a shy, woodland mosquito that has just one or two generations every summer. The tiger mosquito, in contrast, goes through a generation in just twenty days. And it is not shy at all; it follows its human food right into its houses, its backyards, even its sandboxes.

Craig's entreaties have so far fallen soundlessly into the political abyss. In debates over mosquito control, most people fail to get very roused about the threat of arboviral diseases. Arboviral diseases seem so remote; they are rare to begin with, and especially so in Europe and North America where the cold winters keep down mosquito populations. Even in countries where they are endemic, arboviral diseases usually occur in clusters that might be separated by five or ten years; during the intervals, the public's collective memory fades as to how dreadful some of these illnesses can be. Public health officials often run up against a citizenry that is unwilling to be inconvenienced for the sake of avoiding some mosquito bites, and that is unwilling to allow widespread spraying of insecticides when they are not really convinced of the need.

Consider the aftermath of the "flood of the century" that devastated the American Midwest in the summer of 1993. Entomologists were convinced that disease-carrying mosquitoes would have a population surge—especially the mosquito known as *Culex pipiens*, which lays its eggs in polluted waters containing decaying organic matter and raw sewage. *C. pipiens* is a vector for St. Louis encephalitis, and in the wake of the flooding the state of Missouri established a mosquito task force to monitor its population and the proportion of infected mosquitoes and reservoir birds. But even before any worrisome signs were reported in hosts or vectors, residents of Missouri were protesting mosquito spraying programs in prospect. No one understands the risks and benefits enough—sometimes, not even the men and women in charge—to know whether spraying away a potential danger is worth the more proximate danger of the spray itself.

In the summer and fall of 1990, for example, Florida was in the middle of an outbreak of St. Louis encephalitis (SLE), a severe illness that primarily infects the elderly. For those who develop symptoms of encephalitis (a minority of those infected with the SLE virus), the death rate approaches 15 percent. To minimize exposure to the nocturnal *Culex* mosquitoes that carry the virus, the nightly fireworks at Walt Disney World were canceled, swimming pools closed early, and other outdoor evening events were postponed indefinitely. But these changes were accompanied by more grumbling than appreciation. In Vero Beach, for instance, the high school football game—the town's major social event, attracting nearly seven thousand spectators every week out of a total population of just seventeen thousand—was moved up from Friday nights to Friday afternoons. According to the coach's wife, "People were calling up and saying, 'How dare you change the game?' They were acting like their lives were ruined."

In the same way, when homeowners in Massachusetts learned their yards would be sprayed with pesticides in the fall of 1990 to minimize the spread of eastern equine encephalitis, some citizens were furious about disrupting the ecological balance of the surrounding seaside. Conservationists argued that spraying might kill so many of the natural predators of mosquitoes, like dragonflies and minnows, that the mosquito population would actually increase in the long run. State officials, who knew that EEE had already killed one of its three victims that summer, took offense. As one official huffed at a news conference, "We're comparing minnows to human lives? Let's be realistic."

"The public," whoever it may be, has frequently demonstrated its unwillingness to alter comfortable habits for the sake of some greater good. Consider the great struggles that have erupted recently over whether to allow smoking in restaurants and offices, whether to require households to recycle paper and plastic, whether to allow residential construction on sites too close to toxic dumps or Lyme disease vectors or other possible health dangers. This unwillingness shows up again in considering steps required to eliminate certain arboviral diseases. In the western United States, for instance, Colorado tick fever—which is serious but rarely fatal—could be all but wiped out with a few simple changes. The ticks that carry the CTF virus infect ground squirrels, gophers, and other small mammals that live in Colorado, and humans are exposed when they camp in some of the state parks close to where the animals are. "It can all be prevented," says Shope of Yale. "We could kill off all the cute little animals—though nobody would want that—or we could open the parks only during the times of the year when the ticks are not active, and build campsites in places that don't have so many ground squirrels." None of these changes has ever been implemented, though; in Shope's opinion, "People are willing to tolerate getting sick in order to be close to nature."

That is, in essence, the problem when it comes to controlling emerging arboviral diseases. For no other virus are the prodromes as clear—first the vector arrives, then the virus, then the animal epizootic, then the human epidemic—and for no other virus are there as many options in interfering with the inevitable progression toward human illness and death. But without the political will to intervene, all the forewarning is useless. Public officials don't know how to practice true preventive medicine, says George Craig; "they practice reactive medicine. Even when we have all the conditions ripe for the introduction of a new disease, and when we can do something about it in advance, we are told we have to wait until the disease actually materializes."

# 7

# *The Emergence of a New Flu*

Scary viruses with fancy names give glamour to the study of emerging viruses and seem to keep them all at a safe distance. If most new viruses come out of Latin America, Asia, or Africa, people elsewhere in the world feel they can rest a little easier. How can there be any real threat from a virus whose name they can barely spell? But the next major emerging virus, according to many experts, will probably not be a bizarre one at all: it will probably be influenza. And if that sounds like an anticlimax, it's only because we haven't faced a major pandemic strain of influenza for more than twenty years. (The term *pandemic* means a higher-than-expected rate of disease occurring on several continents at once, as opposed to an *epidemic*, in which the excess disease is confined to a particular area.) The worst pandemic in centuries hit in 1918, when between twenty million and forty million people around the world died of influenza. If a similar strain were to emerge today—a strain that, last time around, killed literally overnight—some experts believe that even modern medicine would be helpless to prevent many related deaths.

"Influenza is a horrible, fulminant, and rampant disease," says John La Montagne, chief of infectious diseases at the National Institute of Allergy and Infectious Diseases in Bethesda, Maryland. "I heard one story from the 1918 pandemic about a bridge group, four ladies who were all well enough to play cards together until about eleven o'clock. The next morning, three of them were dead." In October 1918, influenza killed 196,000 people in the United States—almost twice as many in a single month as died of AIDS during the first ten years of that epidemic.

Before the dreadful winter of 1918–19 was over, two billion people around the world had come down with influenza. The pandemic caused more death and dysfunction in one six-month period than in any comparable period, before or since. It was "the most devastating epidemic that we have ever had in history, and it happened in this century," says La Montagne. "No one really knows why it occurred, but there's every expectation that if it occurred once, it can occur again."

We like to believe such plunder is an ancient relic; whatever was killing people so ruthlessly in 1918 must certainly be something we can treat by now. It's true, of course, that modern medicine has given us an influenza vaccine, an anti-influenza drug (amantadine), and plenty of antibiotics to prevent or treat secondary bacterial infections. But during the 1918 debacle, many victims were felled in mid-stride, the way the bridge players were. One man, for instance, got on a streetcar feeling well enough to go to work, rode six blocks, and died. In the face of a virus that kills so rapidly, all the antiviral drugs in the physician's armamentarium would be impotent. And even the influenza vaccine, which must be reformulated each year to keep pace with the newest variants of this fast-mutating virus, would take so long getting manufactured and distributed that thousands might die waiting.

Today, with less-than-pandemic strains of influenza that strike anew every fall and winter, influenza is still a major killer. Its most frequent victims are the elderly and the chronically ill, two groups that are increasing faster than the general population throughout the industrialized world. During a typical influenza outbreak, one-quarter of the population—regardless of age, which is unrelated to susceptibility—will fall ill. And a not insignificant proportion of them, mostly the very old and the chronically ill, will die. Influenza is the sixth leading cause of death in the United States, responsible for some ten thousand to fifty thousand deaths every year. And because it spreads so pervasively, it is also enormously expensive, in terms of both medical expenditures and lost wages. The Centers for Disease Control places the tab for influenza, even during an ordinary flu season, at nearly $10 billion a year in the United States alone. As one influenza researcher has put it, "Flu is not so dramatic a disease, but clearly in terms of numbers over the years, AIDS is peanuts."

Despite these somber statistics, influenza still has a reputation as a relatively benign, self-limiting infection. Maybe it's that pervasive phrase, "just the flu," conveying as it does a certain harmless inevitability. It makes influenza seem like no big deal—sort of the common cold writ large. "Seasonally I go on TV and say, 'Surprise, surprise! We have a flu epidemic!'" says Edwin D. Kilbourne, professor of microbiology at the Mt. Sinai School of Medicine and a national authority on influenza. "There's no interest in the disease whatsoever between epidemics. But then people get all excited and the TV cameras appear when the epidemic comes. It's all very well to warn people about threats, but it's only when the reality emerges that they pay attention. I think the truth of this has been dramatized with AIDS."

An ordinary case of influenza usually takes about seven to ten days to run its course. During that time, a person suffers from high fevers (usually in the range of 102° to 104° F, or 39° to 40° C), for at least the first three or four days, accompanied by many of the complaints of fever: chills, sweating, exhaustion, headache, light sensitivity, muscle aches and pains (especially in the back and legs), and, frequently, coughing, sneezing, sore throat, and chest pains. When a major new pandemic strain erupts, total numbers of cases are higher, and proportionately more deaths occur. And what makes pandemics especially lethal is that they are also often accompanied by life-threatening complications, affecting not just the very old and very young (who are at greatest risk of dying during ordinary flu seasons) but also people who were previously quite well. Pandemic influenza can progress rapidly into a lung infection called primary viral pneumonia. The patient has trouble breathing, turns blue, spits up blood, and suddenly dies—as quickly as forty-eight hours after the first signs of infection.

The cyclical emergence of pandemic influenza offers the opportunity to implement some of the predictive, and possibly preventive, strategies that scientists want to use to head off more exotic viral threats. Influenza has characteristics of many other emerging viruses that make it amenable to "viral traffic control": animal reservoirs that can be monitored for signs of increasing infection; a vaccine that can be tailor-made and administered far more efficiently than it currently is; physicians in far-flung communities who can be enlisted to serve as bellwethers for new viruses; close contact with public health officials internationally—especially, in this case, in China, where pandemic strains have historically originated. This is not simply a dress rehearsal for some bigger, more important disease; pandemic influenza will be, almost without a doubt, a major plague when it emerges, probably in the next several years. The nature of the surface antigens of the current predominant strain of influenza has not changed appreciably since 1968. (The annual slight changes in those antigens, daunting as they are to any individual's immune system, do not lead to global pandemics.) If history is any guide, we can probably expect a major alteration of those antigens—one big enough to lead to a worldwide outbreak of severe flu—before the century turns.

Unlike almost every other emerging virus, the influenza virus has mastered and become dependent on just one terrific trick: the speed with which it evolves. More stable viruses, like smallpox and polio, are rather easy to control with an effective one-time vaccine. Not influenza. Because the virus mutates so frequently, the influenza vaccine is concocted anew each year, based on scientists' best guess of what surface proteins will be coded for by the virus involved in that year's outbreak. (The surface proteins, after all, determine precisely what antibodies will be formed in response to an infection or vaccination. Only one antibody will fit tightly enough to neutralize the virus.) Once they forecast the probable antigenic composition of the virus, scientists try to use those very antigens to make their vaccine, in the hopes that they will stimulate immunity to the virus that is circulating at the time. But they run a high chance of being wrong—some observers have put odds of success at no better than fifty-fifty—and even when they are right the vaccine lasts only as long as that year's strain. By the next flu season, the fickle virus is almost certain to be wearing a new protein coat, a disguise sufficient to slip right past whatever antibodies were formed in response to the previous year's shot.

Because mutation is such a critical component of the virus's reproductive strategy, and because mutation accounts for its designation as an emerging virus, any discussion of influenza must begin with a discussion of how the virus itself changes from one strain to the next. The mutations that lead to major changes in viral surface antigens, the ones that create the pandemic strains of influenza, are quite rare. They have occurred just three times in this century: in 1918, before the "Spanish flu" pandemic; in 1957, when a new strain emerged called the "Asian flu"; and in 1968, when yet a third strain was introduced, known as the "Hong Kong flu."

Compared with the Spanish flu pandemic, the more recent outbreaks were extraordinarily mild. When the Asian flu first emerged—

meaning that no one alive had previously been exposed to precisely that configuration of surface antigens, with the possible exception of people over the age of eighty—the attack rate was the expected 25 percent, but mortality was relatively low; about seventy thousand Americans died. With the Hong Kong flu, there were just twenty-eight thousand deaths. This lower mortality rate could be traced to the fact that of the two major antigens on the virus's surface, the Asian flu and Hong Kong flu differed in only one. This meant that most people had at least partial immunity to the new 1968 virus, which might have tempered its effects.

The explanation for the influenza virus's mutability lies in the arrangement of its genes. Because its genetic material is packaged as RNA rather than DNA, random mutations during replication are relatively common. As with other RNA viruses, when the influenza virus makes a mistake in lining up nucleotides for copying, it skips the step available to DNA viruses to correct errors. So these uncorrected errors, if they are in the genes that code for certain proteins, are passed along to the progeny viruses, leading to minor changes in the virus's surface antigens—changes that are known as antigenic *drift*. Antigenic drift is what renders ineffective the antibodies formed in response to an old influenza infection. It is as though the virus took off its purple coat and put on a red one of the same cut and cloth; the body can recognize the new virus as familiar, but the differences mean that its antipurple techniques will no longer quite be enough.

Antigenic drift occurs every year or so, rendering any prior immunity to influenza, whether derived naturally or through a vaccine, ineffective after a very short time. Someone who received a flu shot in 1992 will have little immunological memory for the slightly changed influenza virus encountered in 1994. (Natural immunity, the result of the body's struggling against an actual bout of influenza, usually lasts longer than vaccine-induced immunity, but it too is finite.) Random mutations—either point mutations (also called substitutions) or deletions—account for most of antigenic drift. Either one of these errors can occur during gene replication, when a single strand of the negative-strand influenza RNA serves as a template (the way messenger RNA does in an animal cell) along which a complementary strand of nucleic acid is to be assembled. In a point mutation, the wrong nucleotide is placed alongside one spot on the mRNA template and somehow gets inserted into the line. In a deletion, no nucleotide is brought forth at all—which means that in reading off the three-nucleotide codes for amino acids later in the replication process, the triplets will be askew and the wrong proteins will be assembled.

Point mutations and deletions among RNA viruses are much more common than scientists once believed. John Holland of the University of California at San Diego recently calculated that a mutation occurs once in every ten thousand viral replications. In other words, for every ten thousand new viruses made—which takes less than an hour in an active infection—one of them will be a mutant. This is six orders of magnitude greater than the rate at which mutation occurs in human cells, meaning that mutations are ten to the sixth power ($10^6$, or one million) times more likely to happen to a virus than to a human cell. At this rate, Holland concluded, it takes just twelve weeks to change a virus's nucleotide arrangement by 15 percent.

In the vast majority of mutations, there is no functional effect on the influenza virus itself. Either the virus continues producing its proteins unimpaired (this is possible because more than one triplet of nucleotides can code for the same amino acid), or the mutant dies. But sometimes—probably as rarely as 1 percent of the time—the new influenza virus survives, is antigenically unique, and is still capable of infecting human cells.

A different, more significant change in the influenza virus can occur because of another peculiarity in the way the virus itself is put together. It has what is known as a segmented genome, a weakly bound string of genes with clefts between each region. There are eight segments, operating much like a primitive arrangement of proto-chromosomes, with each segment responsible for the manufacture of one or two distinct viral proteins. They are physically connected, but only loosely so; they easily come apart and rearrange with other viral segments if different influenza viruses are nearby. The insertion of new segments from different viruses, especially from viruses of animal origin, lead to genetic "reassortants." These reassortants, if they involve genes that code for proteins in the virus's surface antigens, can lead to major changes in the influenza virus's configuration that constitute what is known as antigenic *shift*. Antigenic shift goes far beyond replacing a purple coat with a similar red one. It is more like taking off that purple coat and putting on a white tunic, green scarf, and spangly orange cloak. Everything is different, and the body fails to recognize the virus altogether.

For genetic reassortment to occur, a cell must be simultaneously infected with more than one influenza virus of more than one animal strain. Such co-infection does not readily occur in humans, even though they are susceptible to swine as well as to human influenza, but it happens often in the reservoir animals. These animals, primarily ducks, other birds, and swine, can be infected with influenza from any source—human, avian, or mammalian—without getting sick. Inside a reservoir animal's intestines, viruses flutter about as though they are in the hands of a master card shuffler, moving places randomly and falling back in line in random new combinations. Fortunately for us, though genetic reassortment itself is rather frequent, the creation of a new reassortant that is capable of infecting human beings is quite rare. When such a human-infecting reassortant emerges—part human, part bird or pig—the potential for a new influenza pandemic has arrived.

Of all the reshuffling of gene segments that goes on in the guts of pigs or ducks, only once every ten to forty years does a new virus emerge that can rage through a human population. Most reassortants are not viable, like the result of a chance mating between a chimpanzee and a marmoset. But a hybrid virus that *is* viable, and is capable of infecting a human cell, is so radically different from any previous influenza virus that it can infect large populations quickly and cause serious complications.

Pandemic influenza has historically originated in China—even the misnamed Spanish flu of 1918 has Asian origins—primarily because the country has so many ducks. (Wild ducks are the predominant reservoir for influenza, more so even than horses or swine.) Waterfowl are welcome in China; they prey on many of the pests that would otherwise plague rice crops. Indeed, by some estimates, China has more ducks than it has people. On Chinese farms, ducks live close to humans and close to many of the other farm animals that are also influenza reservoirs; the ducks are as likely to be infected with human influenza as with avian. These co-infections provide the ideal medium for genetic reassortment. More trouble erupts on farms where ducks and chickens are raised in proximity to pigs, another common practice in China. A pig that is co-infected with influenza from different species can serve as a mixing vessel for the creation of entirely new strains.

"This is startling information," says Stephen Morse, referring to the notion that integrated pig-duck farming is responsible for the reassortment of new pandemic influenza strains. Morse helped popularize this theory by inviting one of its leading proponents, Robert Webster of the St. Jude's Children's Research Hospital in Memphis, to speak at his 1989 conference. What makes it so startling is that it postulates influenza, long thought to be the most dramatic example of random mutation in all of virology, as no less subject to the actions of human beings than any other virus. "Influenza has always been described as the classic example of viral evolution at work, and scientists have long believed that new epidemics are caused by mutations in the virus," Morse says. "Although this may be true of the smaller annual or biennial epidemics we frequently experience, it is apparently not true of the influenza viruses that cause pandemics."

Ducks have other habits, besides a tendency toward co-infection, that also make them the primary source of pandemic influenza strains. They are perfectly suited to spreading the reassortants around. During their seasonal migrations, ducks can fly enormous distances, spreading their contaminated feces across large expanses of countryside. And they take in water through the cloaca, or rectum, so that as they swim they are simultaneously drinking pond water tainted with their own or other ducks' virus-filled excrement.

The passage of influenza from duck to duck is quite different from its passage from person to person. For ducks and other animal reservoirs, influenza is confined to the intestinal tract: it comes in through the mouth (and, in ducks, through the cloaca), replicates in the gut, and is released in the feces. In humans, influenza is a respiratory infection. The virus goes into the body through the nose and throat, multiplies in the lungs, and comes out for transmission to the next host via coughs and sneezes. (The virus can remain airborne for as long as two hours.)

"If you could eliminate all the horses, swine, and ducks in the world, you could eliminate pandemic influenza," says Kilbourne of Mt. Sinai. Inside the animal intestine, two influenza strains swapping segments as they replicate could yield a highly pathogenic strain of human influenza. It's more likely, of course, that the reassortant is either harmless or unable to survive, but when enough rearrangements happen over enough replications, the fluke pandemic strain does occasionally emerge. Suppose the new reassortant contains seven segments from the human influenza virus and one segment from the avian. And suppose that single-segment avian gene codes for a new surface antigen, one of those capable of gaining entry to human cells. If all these conditions exist, the new virus could lead to infection in any human who gets too close to the animal's droppings. And that new virus would be so different from all previous strains that no one on earth would be immune.

The last time such a dangerous reassortment occurred was in 1968. But today, the adoption of certain new farming techniques in the developing world actually increases the odds that another such strain will soon emerge. Pigs, ducks, and chickens live side by side on many Asian farms, especially those engaged in the increasingly popular fish farming. Widely promoted as an energy-efficient way to generate high yields of protein foods, fish farming involves feeding hen feces to pigs and fertilizing fishponds—where ducks also swim and drink—with fresh pig manure. Some virologists, especially in Europe, worry about the prime opportunities for genetic reassortment that this agricultural method provides. As scientists from Wales and Germany recently wrote about fish farming, "The result may well be creation of a considerable potential human health hazard by bringing together the two reservoirs" of influenza viruses.

Any discussion of antigenic drift and shift requires an understanding of the two surface antigens of the influenza virus: hemagglutinin (known by the shorthand H) and neuraminidase (known as N). These are the antigens against which antibodies are formed during a primary infection. A minor change in H and N antigens, one that allows them to stay classified as the same type from one year's outbreak to the next, qualifies as antigenic drift. A major change, leading to a recategorization of that year's influenza as either a new H type, a new N type, or both, is antigenic shift.

Antigenic shift has happened only three times this century, leading to the pandemics of 1918, 1957, and 1968. By studying the antibodies still harbored by people who survived these plagues as children, virologists have been able retrospectively to determine precisely what kinds of H and N types were involved each time. These retrospective studies depend for their usefulness on an important observation about influenza antibodies: the antibodies that are most robust, and that last a lifetime, are those developed in response to an individual's first influenza infection during childhood. This holds true even for people who are subsequently exposed to different pandemic strains. The surface antigens involved in the Spanish flu pandemic of 1918 are now designated $H_1N_1$ (type 1 hemagglutinin, type 1 neuraminidase).

FIGURE 6. **Influenza's antigenic drift and shift**

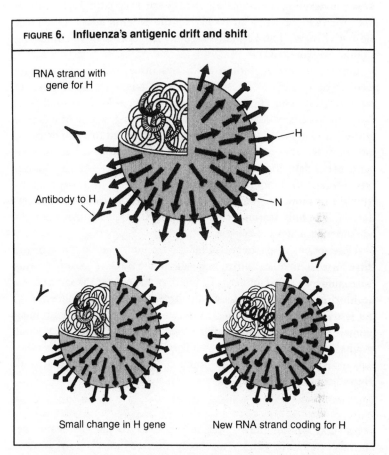

RNA strand with gene for H

H

Antibody to H

N

Small change in H gene

New RNA strand coding for H

Figure 6. Influenza virus is recognized by the body because of its two surface antigens: hemagglutinin (known by the shorthand H) and neuraminidase (known as N). These are the antigens against which antibodies are formed during a primary infection. A minor change in the genes that code for H and N antigens allows them to stay classified as the same type from one year's outbreak to the next. This results in a small change known as antigenic drift (bottom left). If an entirely new gene segment appears, the surface antigens change sufficiently to be re-categorized as either a new H type, a new N type, or both. This large change, which accounts for new pandemics of influenza, is antigenic shift (bottom right).

Those involved in the Asian flu of 1957 are designated H2N2; those from 1968's Hong Kong flu are H3N2 (the neuraminidase component did not change). The H3N2 configuration is still the one circulating around the world today.

Only three hemagglutinin types and three neuraminidase types seem to be capable of infecting humans. (These are hemagglutinin types H1, H2, and H3, and neuraminidase types N1, N2, and N8.) But there are many other antigen types circulating, a total of fourteen H types and nine N types; most of them are limited to avian or swine influenza. Recent research has indicated, though, that at least one antigen previously thought unable to infect humans—hemagglutinin type 7—can infect monkey cells in the laboratory. This raises the possibility that more antigens than we realize can turn into pandemic influenza. The only reason we have not seen it happen is that the right reassortant has not yet emerged.

Descriptions of pandemic influenza are limited to just one of the three forms of influenza virus, known as influenza A. Two other forms exist, influenza B and influenza C, but they do not pose the same public health risk. Influenza B causes much the same illness as influenza A, but it is far more stable, primarily because it cannot infect nonhuman animals that can serve as mixing vessels for creating new pandemic strains. And influenza C, though it is related structurally to the other two types, causes only a mild upper respiratory infection that is all but indistinguishable from the common cold.


For a short while, influenza researchers believed that pandemics occurred every eleven years. This theory, derived in the late 1970s when it looked as though swine flu would be the next pandemic, was based on a cycle of three large-scale outbreaks: 1946, 1957, and 1968. Only now do we realize, in retrospect, that the 1946 outbreak represented not a totally new antigenic shift, but a higher-than-expected incidence of an old influenza.

The eleven-year truism has gripped the public imagination, however—and captured the fancy of a few scientists as well. Back in 1976, on the very day that the nation's first case of swine flu was being treated in New Jersey, an op-ed piece coincidentally appeared on the pages of the *New York Times*. The article was written by Edwin Kilbourne, who wanted to encourage all Americans to get vaccinated against influenza that winter. His article was prophetic of the emergence of a new virus, but his timing was off. He began his article this way: "World-wide epidemics, or pandemics, of influenza have marked the end of every decade since the 1940s—at intervals of exactly eleven years."

Even today, the obligatory newspaper article that appears every flu season seems always to contain some reference to the "fact" that influenza pandemics occur every eleven years. But the popular press notwithstanding, most contemporary influenza researchers, Kilbourne included, say the notion that pandemics occur every eleven years is simply wrong. "It's a misinterpretation of old data," says Brian Murphy, chief of the respiratory virus laboratory at the National Institute of Allergy and Infectious Diseases, "generated before the origins of epidemics became clear, or before sequence analysis on the virus genome was possible." About a generation ago, blips of increased influenza activity did seem to take place at roughly eleven-year intervals: in 1946, 1957, 1968, and 1977. But now we know that in 1946, the virus in circulation was still the same H1N1 that had been in existence since 1918. And now we know that, in 1977, there was a perfectly logical—though rather bizarre—explanation for what happened that had nothing to do with antigenic shift. When the virus from that year's outbreak, nicknamed the "Russian flu," was examined genetically, it proved to be another H1N1 variety. Curiously, though, further molecular analysis revealed that the Russian flu virus was genetically identical to a type of influenza virus that had last been seen in the early 1950s, more than twenty-five years before. "You can anticipate certain rates of change in each gene per year, even if the hemagglutinin and neuraminidase stay the same," Murphy says. "But this virus looked like it was in a genetically frozen state." And by "genetically frozen," Murphy means, literally, just that: frozen. Many virologists now believe that an influenza virus from the 1950s, stored in a laboratory freezer at a research facility in China, somehow got released into the environment in 1977. Murphy realizes this is an undiplomatic thing to say, since it implies sloppiness on the part of the Chinese investigators, but no one has ever come up with a more likely explanation. The accidentally unfrozen virus has since continued to circulate and to alter its genome gradually from year to year, the way it would be expected to. Now, two different influenza types coexist around the world: the Russian flu and the Hong Kong flu, types H1N1 and H3N2. Some years one predominates, some years the other does.

But old myths die hard, and sometimes they lead to new myths of their own. In 1990, the British astronomer Fred Hoyle—who brought us the theory that all viruses, indeed all life-forms, have fallen from space—used that outmoded eleven-year-cycle theory of influenza to add weight to a theory of his own: that earthly diseases are intimately tied to intragalactic events. He and his colleague, mathematician N. Chandra Wickramasinghe, wrote that for the last seventeen cycles, pandemics here at home "appear to have kept in step" with a synchronous cycle of maximum sunspot activity. They traced the association to the intense solar winds that sunspots generate. The winds "can rapidly drive charged particles of the size of viruses down through the exposed upper atmosphere into the shelter of the lower atmosphere." Once these viruses settle to the earth, wrote Hoyle and Wickramasinghe, they spread wildly from one person to another, and a new influenza strain emerges.

But while the eleven-year cycle may be a myth, there are good reasons for suspecting that influenza emerges in a cycle of some kind—one that suggests that the next pandemic may be caused by an $H_1$ surface antigen. The three H types that infect humans seem to take turns, in order, as the pandemic strains change. $H_1$ seems always to beget $H_2$, $H_2$ to beget $H_3$, $H_3$ to beget $H_1$ again. It seems to work this way even retrospectively, when scientists try to extrapolate, from the blood of older individuals, what type of antigens were responsible for pre-1918 pandemics. In the mid-1970s, researchers analyzed the antibodies still circulating in the bloodstreams of very old people who had been exposed to influenza pandemics in the late nineteenth century. They found that people in their nineties still had antibodies to $H_2N_2$—the probable cause of a pandemic from the mid-1800s—and those in their seventies and eighties had antibodies to a different variety, $H_3N_2$—the probable cause of the pandemic of 1890. (Another clue about the strain responsible for earlier epidemics was the fact that in 1968, when the $H_3N_2$ Hong Kong flu emerged, people over age seventy-eight—who had been children during the 1890 pandemic—had lower mortality and morbidity rates than people who were ten or fifteen years younger.) If the H types really do cycle in this way, the next one in line to emerge is type $H_1$, the same type responsible for the rampant devastation of 1918.

In evaluating the mutability of the influenza virus, it seems logical, almost inevitable, to wonder why it mutates as often as it does. But another question, just as pertinent, would be this: Why do pandemic strains arise so rarely? The big antigenic shifts, after all, are not that common; forty years passed between the Spanish flu and the Asian flu, and more than twenty years have passed since the H3N2 strain emerged in 1968. The rarity of these shifts was the topic of Stephen Morse's opening address before his 1989 conference on emerging viruses. "What stabilizes viral 'species'?" he asked that morning in May. "Why doesn't influenza A turn into influenza C? Every virus makes thousands or millions of progeny in each infection and—given the high mutation rates—many of these must be variants. And yet only relatively rarely do we see successful new types emerging." It is probably a lucky stroke for people that new influenza chimeras are so vulnerable to die-off when they hit the atmosphere. Otherwise, every time a new influenza recombinant emerged from the guts of ducks, we might have trouble indeed.


A population grown complacent about the threat of influenza received a stunning shock in the mid-1970s with the explosion of a childhood disease called Reye's syndrome. In 1980, 555 American children were diagnosed with Reye's syndrome, a number that reflected a slow and steady increase over the previous few years. (The Centers for Disease Control began surveillance for Reye's syndrome in 1976.) The children suffered from profuse and continuous vomiting, personality changes, progressive lethargy, and, in severe cases, coma. Twenty-three percent of them died that year; in other years, the case-fatality rate sometimes exceeded 40 percent. Clearly, something new was happening. And it seemed related to something dreadfully familiar: a bout of influenza or of some other common childhood infection, particularly chickenpox. For the first time in generations, American parents felt influenza was something to fear.

Reye's syndrome was first described in 1962, when Douglas Reye (his name rhymes with *eye*) noticed a bizarre set of symptoms among twenty-one young patients at the Royal Alexandra Hospital for Children in New South Wales, Australia. In the seventeen children who died, Reye, a pathologist, found characteristic anatomical changes that almost looked as though these children had been poisoned. They had swollen brains, damaged kidneys, and highly unusual livers: slightly enlarged, firm, and bright yellow in color. The following year, an

American physician, George Johnson, noticed similar changes among sixteen children who died of encephalitis during a four-month influenza outbreak in a small town in North Carolina. On autopsy, these children looked much the way Reye's patients had, and Johnson concluded that the syndrome was somehow a complication of their exposure to influenza. Thus was born the germ of an understanding of Reye's syndrome—which is officially designated Reye-Johnson syndrome.

But why did some children develop Reye's syndrome after influenza and others not? That question could be answered only with case-control epidemiological investigations. In the late 1970s, public health officials in Arizona, Michigan, and Ohio compared children who developed Reye's syndrome following respiratory viral infections to children who had the same infections, at about the same time, but who did not develop Reye's. They asked both sets of parents everything that had happened to the children while they were sick: what they ate, how high their fevers got, what medications they took. And the only difference between the Reye's group and the controls was that the children who developed complications had, during their original illnesses, taken aspirin.

By 1982, the etiologic link between aspirin and Reye's syndrome was well established. The surgeon general of the United States issued a warning against giving aspirin to children with flu or other viral infections. In 1988, that warning became law: the Food and Drug Administration required that aspirin bottles feature a prominent label stating that aspirin should not be used by children or teenagers who have symptoms of chickenpox or influenza. "The association between aspirin and Reye's syndrome is so strong," says Lawrence Schonberger, a CDC epidemiologist, "that it has now become literally foolhardy to act as if no etiologic relationship exists."

But while the actual relationship between aspirin use during viral infection and subsequent Reye's syndrome is generally accepted, the explanation for that relationship is not. The connection is far from a perfect one: of all the children and teenagers who *do* take aspirin while sick with influenza or chickenpox, federal officials estimate that no more than 0.1 percent of them develop Reye's syndrome. And a nagging 5 percent or 10 percent of children who *do* get Reye's never had a viral infection. But as the incidence wanes—in 1990, just twenty-five American children developed Reye's syndrome—scientists seem unwilling to explore this mystery further. So the complication will join the ranks of other rare complications from influenza. The virus has, albeit rarely, been shown to invade cells of the central nervous system or muscle, leading to infections there—known, respectively, as encephalopathy and myositis—with occasional long-term residual effects. But this is little different from the behavior of any other acute virus infection. Sometimes, on its way out of the body, the virus simply gets derailed.

In one of the satisfying twists that occasionally surface in virology labs, researchers engaged in influenza vaccine development have managed to outsmart their changeable little foe by turning its own best weapon against it. The laboratory strains developed for the influenza vaccine, the ones that can stop an epidemic in its tracks, have capitalized on the virus's notorious ability to recombine—the very trait, ironically, that makes that same vaccine so short-lived. The recipe is simple. Take that year's variant of the influenza virus, throw it into a stew with a lab donor strain of virus that leads to rapid proliferation, and let the virus do what it does best: incorporate that fast-growing strain into its own genome and start replicating like mad. From there, it's an easy matter to take those plentiful viruses and attenuate them for a flu vaccine.

Here is where the nice ironic twist occurs. The virus's promiscuity originally led to the culturing, in the 1930s, of an influenza strain known as PR-8, which contains the particular combination of genes needed to adopt an extraordinary growing habit. When a sample of any year's new influenza is cultured with PR-8, after a certain number of passes some of the new influenza picks up those growth genes from the PR-8 culture, and the new reassortant virus proliferates rapidly enough to make a new vaccine.

Fertilized eggs are the medium on which influenza viruses are cultured for vaccine production. Into each egg, laboratory workers mix some PR-8, a bit of that year's influenza variant, and, on the next passage, enough serum to keep the PR-8 from growing too fast. The serum allows the influenza to pick up the growth genes from the PR-8 while still retaining its own genes for its surface antigens.

Vaccine manufacturers must know by mid-February of any year which antigens to include in the following year's formulation. In other words, while the nation is still in the grip of the current influenza season, scientists must determine what virus will cause next year's illness. Pharmaceutical companies need that long a lead time to ensure production of adequate amounts of vaccine ready for October delivery, so physicians can begin administering it by November. The vaccine does not work immediately; people usually require two weeks or so to make enough antibodies to confer lasting immunity. This timetable allows readiness for influenza season, which usually begins in December and lasts about twelve weeks, until the end of February.

The production schedule was speeded up considerably during what turned out to be one of the most controversial public health campaigns in recent memory: the swine flu mass immunization program of 1976. The large-scale effort to manufacture and distribute enough vaccine for "every man, woman, and child in the United States" (in the words of the president, Gerald Ford) showed that millions of doses of high-quality vaccine could be made in a matter of three or four months. But because the dreaded pandemic never materialized, some critics have said in retrospect that forty million Americans were vaccinated for nothing. In fact, the only real illness to result from the swine flu adventure was actually caused by the vaccine: about one thousand people developed Guillain-Barré syndrome, a serious paralytic disease that could be traced directly to an immunological response to the inoculation. Still, Edwin Kilbourne—an early and consistent advocate of mass immunization—says that based on what was known at the time, the government acted rationally and prudently. Scientists thought they had encountered the same influenza strain that caused the devastation of 1918, and they thought they had a tool available to help them avert hundreds of thousands of deaths. "Better a vaccine without an epidemic," Kilbourne says today, "than an epidemic without a vaccine."

Prudent or not, the government's immunization campaign turned out to have some long-lasting political repercussions. David Sencer, the director of the CDC, was driven out of office for the part he played in arguing that even if swine flu never materialized, it was better to err on the side of caution. In a now-famous memo, Sencer pushed for universal immunization in the face of uncertainty, pointing out that "the Administration can tolerate unnecessary health expenditures better than unnecessary death and illness." Some political pundits even blame the program, at least in part, for President Ford's defeat in the elections that November, one month after vaccinations began and one month before they were aborted.

The story really began months earlier, in February 1976, on a Friday the thirteenth. Reports came in to CDC of an influenza outbreak at Fort Dix, a boot camp in New Jersey. One recruit, nineteen-year-old Pvt. David Lewis, died. Of nineteen specimens examined for virus typing, five—including the one from Lewis—were shown to have the same $H_1N_1$ arrangement as the influenza virus thought to be responsible for the 1918 pandemic. This got many public health officials into gear, hoping to use modern technology, namely vaccination, to head off a disaster of the magnitude of 1918. But in hindsight, those involved in the decision making have learned that major variants of influenza virus in a few individuals do not necessarily signal the start of a new pandemic. "Early detection of a new virus may not be adequate evidence on which to undertake mass immunization," Kilbourne wrote in a 1979 article called "The Virus That Vanished." "But it is, I believe, a signal at least to produce vaccine to hold in readiness."

Ironically, the swine flu experience might have uncovered a hidden hazard of surveillance: by looking too hard, you might actually turn up something that you just cannot interpret. When influenza broke out at Fort Dix, an aggressive young public health director, a man coincidentally trained by Kilbourne, was called in to identify the responsible strain. He had learned his lessons well: even after he examined the first several samples of respiratory tract washings and found nothing unusual, he continued to look at more. "If he had not gotten eighteen or nineteen throat washings, if he had stopped at two or three, then on a probability basis he would have decided the epidemic was caused by $H_3N_2$ influenza, and nothing would have happened," says Kilbourne now. "No action would have been taken, because no threat would have been perceived."

Such active surveillance as led to the "discovery" of swine flu continues to this day. One hundred sixty U.S. internists are recruited annually as CDC "sentinel physicians," sending a weekly postcard to the headquarters in Atlanta to let government officials know how many patients nationwide are complaining of flulike symptoms. A subset of these doctors also sends sputum samples to a central government laboratory, which types the influenza virus and phones in the results to CDC. More elaborate scouting has involved annual expeditions to the countryside of China at the start of every flu season. (In southern China, flu season is June through August, in synch with the southern hemisphere's; in northern China, it is in synch with the rest of the northern hemisphere's flu season, December through February.) Because influenza strains usually emerge from China, these expeditions have often paid off. In late 1987, for instance, a CDC trip to China yielded samples of a new influenza strain cultured at the Shanghai Hygiene and Anti-Epidemic Center. Antigens from that strain, named A/Shanghai/11/87, could then be incorporated into the U.S. vaccine administered in 1988.

Long a subject of intensive, worldwide surveillance, influenza could offer a blueprint for how other such systems can stay one step ahead of a variety of emerging viruses. But the blueprint would come with a checkered history; with influenza, surveillance has been of only limited success. The virus changes so frequently, and so unpredictably, that the clear-cut clinical benefit of surveillance—an accurate vaccine developed in a timely manner—is often just out of reach. "I am an optimist by nature, but I'm not too optimistic about what surveillance systems can do if they're trying to keep track of many viruses, especially new ones," says Kilbourne. "Influenza watchers, focused on only a single disease, are aware of the lag between discovery of change and implementation of control, and know that only the probability—not the nature—of change can be predicted. Surveillance for the unknown will be even more difficult and will demand a level of clinical and laboratory competence not widely available in the Third World or, for that matter, anywhere else."


Thoughts about surveillance are the inevitable next step when virologists start asking themselves, "What can we do to stop it?" With very few antiviral drugs at their disposal, and with only a handful of vaccines, virologists must feel the way bacteriologists did a century ago: they know what causes human suffering, they know in many cases how to anticipate that suffering, but they are relatively helpless to prevent it. The next wave in biology, the desired end result of analyzing how viruses emerge, will be characterized by the scientists' fighting back.

# PART THREE

# FIGHTING BACK

# 8

# *Anticipating the Next AIDS*

The state of Florida maintains a flock of chickens in each of fourteen counties. Every week during the spring and summer, scientists from the Department of Health and Rehabilitative Services visit
those chickens and take some of their blood. It's not easy to take blood
from a chicken; struggling with these birds can be a mess of feathers,
claws, and clucks. Once the chicken blood is drawn, the scientists take
it back to their laboratory in Sarasota and look for signs of virus infection. They are looking especially for the viruses that cause two
mosquito-borne brain infections: St. Louis encephalitis (SLE) and
eastern equine encephalitis (EEE). Chickens can get either infection,
and so can human beings.

In early June 1990, every one of the chickens in the Indian River
County flock had seroconverted—that is, had gone from having no
antibodies to SLE or EEE the previous week to having SLE antibodies
now. The time factor here was critical, because it showed that the
infection was a new one. In other counties—Lee, Manatee, Orange—
about one-third of the chickens had also seroconverted. The smaller
proportion did not mean they were any less likely to contract the virus;
it just meant they were a few weeks behind Indian River County in
their exposure. By August 13, the other counties had all but caught
up: 80 percent of the chickens in Orange County, for instance, showed
antibodies to SLE.

These chickens were not just any chickens. They were sentinel
chickens. They were part of an ambitious state-wide epidemiology
program begun in Florida in 1982 to anticipate human outbreaks of
viral encephalitis. Like the canary in the coal mine whose death warned

miners there was a dangerous gas leak nearby, a sentinel chicken demonstrates, through the state of its own health, whether it is safe for humans to go outside.

Of course, a sentinel chicken program won't work if the only creatures being monitored are chickens; some sort of human surveillance must be undertaken too. In the counties where the chickens were seroconverting, Florida public health officials routinely monitored hospital records for suspicious symptoms. An increase in viral antibodies in chickens, while leading to this heightened surveillance, rarely seemed to translate into an increase in human cases of encephalitis. But the surveillance system remained in place, and in 1990 it paid off. Among patients hospitalized for encephalitis, five were confirmed to have SLE, and a sixth was presumed to. Knowing what to look for made it easier for physicians to make these diagnoses—and made it easier for public health officials to receive support for a timely antimosquito spraying campaign like the one mounted in July 1993, near Walt Disney World, after six sentinel chickens showed high antibody levels for EEE.

An early-warning system of some kind is our best hope for heading off the next viral plague. The system need not involve chickens, though sentinel animals prone to the same infections as humans are one handy way to monitor the path of a new disease. It need not be located in proximity to the people whose health is being protected; indeed, the best such systems would place sentinel animals, or other relevant methods of early detection of new viruses, quite far from home. And it need not be expensive; estimates are that a good system of surveillance outposts throughout the developing world, strategically placed near mega-cities blooming at the edges of tropical rain forests—the very sites from which new viruses historically emerge—would cost no more to maintain each year than it once cost to inoculate individuals against smallpox.

The invocation of the smallpox eradication campaign in this context is no accident. One of the most vocal proponents of an international network of "virus-tracking centers" is Donald ("D. A.") Henderson, who made his reputation as medical director of the global smallpox eradication campaign. Henderson is now associate director for life sciences at the White House Office of Science and Technology Policy (OSTP), which means, at least in theory, that he has direct access to the president of the United States on matters of public health. At Stephen Morse's 1989 conference on emerging viruses (which took place about nine months before Henderson was appointed to OSTP), Henderson proposed that the United States establish "listening posts" at the edges of rain forests and at other locations from which new viruses might emerge. On site at each unit, he said, should be a diagnostic and research laboratory, an epidemiological team to lead local health care workers in response to an outbreak, and a general clinic to provide care not only for exotic diseases but also for more mundane infections. In addition, each center should have a formal relationship to an American medical school, allowing American students a chance to rotate through the center at some point in their medical training. (Native medical students would receive training there too.) And a central government laboratory located in the United States would be available to coordinate data collection and analysis, offer backup services, and initiate a global response if the situation warrants it. Such a system—involving fifteen overseas centers, ten backup laboratories in the United States, and the government coordinating center—would probably cost some $150 million a year, Henderson said; in comparison, the cost of caring for AIDS patients in a single year will, by one estimate, reach $10.4 billion in the United States alone by 1994.

Henderson's listening posts are an appealing blend of altruism and self-interest. They can be expected not only to improve the health of individuals in the developing nations in which they are located but also, in a roundabout way, to improve the health of people in the United States. According to Henderson, the current failure to teach American doctors-in-training about tropical diseases could well put the future health of all Americans, indeed of all citizens of the industrialized world, in jeopardy. "Young doctors are not acquainted with a number of diseases that we're seeing more of in America," he says. The result of their confusion is that it takes longer to get an accurate diagnosis when new pathogens erupt in unusual locales, leading to a loss of precious time during which epidemics that might have been nipped early are instead granted access to an entire population. If North American and European medical students are given the chance to work at international infectious disease centers as part of their training, then at least a few young men and women will be able to act as resources for others when they continue their professional pursuits back home. "This would ensure," he says, "that any good academic center or large hospital would have a few people who are fully knowledgeable of infectious diseases."

Training in the recognition of bizarre new diseases is more central to the stemming of new plagues than might at first seem apparent. Traditional medical education hammers into the young doctor the importance of playing the odds, of making a presumptive diagnosis of a common condition, not an exotic "zebra"—physician's slang for a bizarre diagnosis, derived from that medical student aphorism about hoofbeats in Central Park. But as the world shrinks and the international exchange of microbes increases, the odds are likely to change, and without warning it might become more likely to find zebras than horses trotting through midtown Manhattan. Tomorrow's physician will have to know how to respond to such sudden shifts in disease patterns—because if they don't, the wheels of epidemic control will not get started. The doctor is the little marble that gets the entire Rube Goldberg contraption in gear; if he or she doesn't sound the first alarm, nothing happens.

"Prompt recognition of an epidemic situation is directly dependent on a high index of suspicion," wrote Wilbur Downs, at the time director of the Yale Arbovirus Research Unit (YARU). "Individual clinicians, epidemiologists, and health services vary greatly in the degree of alertness maintained for possible epidemic incidents." Downs said it was not unusual for "dozens, hundreds, or even thousands of cases" of mosquito-borne illnesses to pass through a hospital's corridors before the first accurate diagnosis is made—thereby allowing millions of virus-bearing mosquitoes to continue breeding and biting, and subjecting individuals to a disease that could have been quickly eradicated weeks earlier with the right diagnosis. Sometimes the delay is due to the physician's sloppiness; sometimes to inattention; sometimes to simple lack of information about the symptoms caused by an exotic disease that is, by definition, encountered only rarely. If the clinician does not know how to recognize early cases of strange new diseases, all subsequent efforts at surveillance and prevention will eventually disintegrate.

Sometimes, however, the failure to call an alert takes place one step up from the clinician, at the level of a ministry of public health. When this occurs, you can bet the motivation is far more venal: the result of a nation's unwillingness to let other countries know about the diseases it harbors. As Downs put it, "Information about occurrence of epidemics may be suppressed in the interest of a lucrative tourist trade." Suppress information about an endemic viral outbreak for the sake of attracting tourists? Ethically deplorable, perhaps, but it happens all the time. It even happens with the knowledge and cooperation of guardians of the public health, not only those from the country in which the disease is occurring—who have an obvious vested interest in denying the truth—but also those from the country actually sending the hapless tourists, in particular scientists from the United States. "If we diagnose an outbreak of something in an area where tourists go and the [foreign] government doesn't want us to announce it, we won't," says Robert Shope, Downs's successor as head of the YARU. (Downs died in 1991.) "We'll report it to the World Health Organization, but it's not our responsibility to go any further; it's that government's responsibility." Shope admits that such a policy is not fair to travelers, but he believes the issue of confidentiality parallels the debate that has gone on recently about the confidentiality of AIDS testing. "Conceptually it's the same argument. The rationale for confidentiality in HIV testing is that without it, people wouldn't go for testing—and that would be even more dangerous. I think the same goes here. If our center didn't promise confidentiality for these countries, they wouldn't send us the samples." In light of this, Shope's advice is quite simple. To people traveling to countries where serious diseases such as yellow fever are endemic, he says: if a vaccine is available, get it.

Uncertainty about how much to publicize adverse findings could well plague a more formal "listening post" network as well, especially if WHO is expected to take the lead. As a loose federation of member nations supported through the United Nations, WHO is acknowledged to be good at building consensus but weak at forcing its membership to see far beyond their competing self-interests. "The fundamental problem with an organization like WHO, as with any part of the U.N. system, is that it's basically a collection of nation-states," says Jonathan Mann, head of WHO's Global AIDS Program from 1986 to 1990. "And so if nation-state A doesn't want the world to know about its cholera epidemic, then WHO really can't do much until or unless the national authorities say, 'Okay, now you can come in and help us investigate.'" In theory, says Mann, WHO should be capable of eliciting meaningful cooperation from countries all over the world; but in practice, this does not always happen. "I worked at WHO and then I resigned, and quitting never affected my judgment of what WHO is capable of doing. It's capable of an enormous amount, but it is not necessarily able to do it. It has to be motivated; it has to be in some ways even forced to assume certain responsibilities."

John Woodall, a scientist in WHO's Division of Epidemiology and Health Situation and Trend Assessment, based in Geneva, has a good deal more faith in his organization. "WHO is the logical international agency to coordinate [viral surveillance] activities," he writes, "since it is acceptable to all countries and able to broker collaboration even between nations that have no diplomatic relations with each other." Woodall notes that WHO has the unique ability to award grants to laboratories in the developing world for the collection and identification of new viruses, much the way Rockefeller Foundation virology labs worked back in the 1950s; to monitor changes in the population of certain virus vector species in various countries; and to engage in disease surveillance in nations bordering those in which an outbreak is occurring, as a way to chart the speed and direction of a virus's spread.

Woodall's defense of WHO came in response to an article, in the fall 1990 issue of a National Academy of Sciences publication, in which Morse laid out suggestions for, in his words, "regulating viral traffic." Among the suggestions were Henderson's "listening post" idea, and a few of Morse's own: "viral traffic planning" as an adjunct to whatever development plans are considered in the Third World; coordination of U.S. viral surveillance at the national level under one umbrella agency (rather than the current patchwork system that involves four or five, with no one explicitly in charge); and the promotion of simple preventive measures, such as the elimination in people's yards of puddles where mosquitoes breed, through public education campaigns. "Emerging disease is just one more consequence of ecological damage," he wrote. "But addressing the problem could provide common ground for uniting otherwise diverse interests—environmental, agricultural, economic, and health—to simultaneously address a wide range of other concerns as well."

Morse was one of the first scientists to start thinking systematically along these lines, but he now has some powerful, and rather vocal, colleagues making similar declarations. Joshua Lederberg, who co-chaired the blue-ribbon panel assembled by the National Academy of Sciences to recommend ways to anticipate the next AIDS (Robert Shope was the other co-chairman), has taken to writing articles and giving speeches on this very topic. "Our preoccupation with AIDS should not obscure the multiplicity of infectious diseases that threaten our future," he has written. "It is none too soon to start a systematic watch for other new viruses before they become so irrevocably lodged."

Not every new disease has to take people by surprise. Certain patterns of viral emergence can be plotted and therefore predicted, to allow for a more pro-active approach to surveillance and prevention. That is why the accumulation of case histories, which especially captivates Morse, has proved so valuable. Lessons can be learned from past mistakes, many of which can be characterized by one striking similarity: the unanticipated hazard of tampering with the environment in the name of progress. The emergence in the 1950s of Argentinian hemorrhagic fever, for example, coincided with the clearing of the grasslands along the Argentinian pampas to allow first-time farmers to plant cornfields. The Argentinian hemorrhagic fever virus, called the Junin virus, is carried by small South American rodents, which found a hospitable new habitat in the rows of maize sprouting up right alongside people's homes. The increase in cases of Argentinian hemorrhagic fever—with fevers up to 104°F (40°C), muscle aches, eye pain, swollen glands, and in some cases bleeding from the gums and nose, blood in the urine, and bloody vomit—could be traced directly to the population increase, and proximity, of infected animals. So these days, when similar agricultural plans are being considered, public health officials should know from experience to test the local rodents for the presence of Junin virus in their bloodstreams. If Junin virus is endemic in the area, control measures can be undertaken to limit the exposure of people to carrier rodents. Development plans can thus be tied to what Morse calls "viral impact assessments," much as they are now tied in U.S. legislation to environmental impact assessments, to estimate the potential effect of changes in the ecosystem on proximate human health.

Sadly, though, foreknowledge alone is not enough; it must be backed up with some course of action. Even if officials are warned, for instance, that local rodents do indeed carry Junin virus, and that the number of infected animals will increase if maize is planted in the regions the rodents traditionally avoid, would that keep officials from implementing their intended policy? Is the temporary prostration of thousands of local citizens (who might get very sick from Argentinian hemorrhagic fever, but are unlikely to die) reason enough to disrupt an important chance at agricultural modernization? Politicians must take into account competing interests, of which citizens' health is only one; others include economic development, national security, balance of trade, tourism. Planting corn might well create a new habitat for virus-bearing rodents; building a dam might well allow for increased hatching of virus-carrying mosquitoes. But how many new cases of disease is it worth to enable a country to mount the agricultural and industrial growth that the corn planting or the dam building allows?

The push and pull of competing interests helps explain the behavior of the nations of West Africa in 1977—behavior that scientists at the time had considered rather befuddling. Wil Downs of Yale was asked to bring a team of American scientists to Senegal as consultants to the United States Agency for International Development (AID), which was overseeing the construction of a series of dams along the Senegal River. AID officials wanted to know whether building those dams entailed any possible health risks to the native populations. This was long before the project was under way—indeed, the dams were not completed until 1985—and there was plenty of time to change direction if the public health implications warranted it. Downs and his associates collected blood specimens from one thousand people who lived along both banks of the river, which divides Senegal from Mauritania, and sent the samples back to New Haven to be analyzed.

"One of the things we found was antibodies to Rift Valley fever," says Shope, who tested the specimens at Yale. "We found it in a pattern which indicated that it was an endemic infection: people were getting infected sort of randomly year after year, yet there were no reports of epidemics, or even of a single case of Rift Valley fever, in either the people or the cattle." The Rift Valley fever virus was apparently being passed around among the residents of the region, but it was causing diseases so mild that no one even recognized it as a clinical syndrome. Because the virus was transmitted by mosquitoes whose eggs developed in standing water, the scientists realized, any change in the amount of water in the region—such as construction of dams—could well result in an increase in the mosquito population, in the number of infected mosquitoes, and in the likelihood of people's coming down with new, potentially more virluent cases of Rift Valley fever.

In his report to the Agency for International Development, Downs said that Rift Valley fever was one of the diseases for which there should be surveillance when the dam was completed, because, as he put it, "you're going to get mosquitoes." But the agency chose to ignore the warning. The dam project continued, and no surveillance was instituted. In fact, the warning itself was totally forgotten, since personnel at the agency shifted several times between the date of the prophecy and the date of its resolution. In 1987, just one year after the Diana Dam was completed, there was indeed an outbreak of Rift Valley fever in Mauritania. It was estimated that more than 1,000 people were infected, and 224 died. "If they had known early on what they were looking for, when they saw the first patient they could have had some sort of mosquito control, and probably could have prevented it," says Shope. "But they didn't know what they had at first; they thought it was yellow fever."

Surveillance for Rift Valley fever could have been mounted rather effectively. Since the disease also affects cattle, a good surveillance system might have involved some sentinel cows; since it is carried by mosquitoes, seasonal trapping and analysis of mosquitoes for Rift Valley fever virus would have helped scientists anticipate the likelihood of humans' getting sick. Indeed, in the years since the Mauritanian outbreak, more systematic surveillance for Rift Valley fever has in fact been instituted. The U.S. Army, for instance, regularly takes infrared photographs of rainfall patterns in the African regions where the disease is endemic. They have shown that periods of high rainfall are associated with high rates of Rift Valley fever—a first step in mounting a mosquito-abatement project during rainy seasons to prevent transmission of the virus.

Other viral diseases, because of the way they are transmitted, are similarly amenable to ongoing surveillance. One of the easiest to anticipate is Bolivian hemorrhagic fever, a disease with much the same symptoms, and much the same epidemiology, as the hemorrhagic fever that affects neighboring Argentina. This often-fatal illness is caused by the Machupo virus, which is carried by a small mouselike rodent known as *Calomys callosus*. The rodent is ideally adapted to domestic life; it likes to live in people's houses and gardens far more than it likes living in open fields. And in Bolivia in the early 1950s, the environment became, almost overnight, especially hospitable to the little *Calomys*. At the time, the plains of the eastern region of Bolivia, in Beni department, were dry, unpopulated, and unspoiled. As Karl Johnson, an expert in Bolivian hemorrhagic fever, recalls, the plains were home to "thousands of stringy beef cattle" owned by a Brazilian family operation called Casa Suárez. The company, Johnson writes, "had German meat processing facilities and a fleet of ships that took beef down the Amazon river system and then to Europe and the Americas. These ships brought back the rice, maize, beans and fruit to feed the cowboys of the Beni."

After the Bolivian revolution of 1952, the new government reclaimed the land from the outside interests that controlled it. On the Beni, this meant that Casa Suárez retreated—as did their boats, which had brought the European foodstuffs on which the Bolivians had depended. The cowboys were, for the first time, forced to fend for themselves. The grasslands proved a poor substrate for farming, so they moved on to the higher *alturas*, with more fertile soil and a better rainfall pattern. It was on these *alturas* that *Calomys callosus* lived—and, with the sudden appearance of homes and gardens nearby, this is where the rodent and its virus thrived. By 1960, when the population of the Beni had shifted to the *alturas*, outbreaks of a new hemorrhagic fever were first reported. It caused very high fevers (ranging from 102° to 105°F, or 39° to 40.5°C), bleeding from the gums, nose, stomach, intestines, and uterus, tremors of the tongue and hands, profound weakness, and hair loss.

But by 1964, when outbreaks reached a high of nearly seven hundred cases in a single year, the nation was already beating the virus into retreat. Bolivia has had fewer than one hundred cases in all the years since 1964. The reason for the decline: intense and continuous surveillance. Beginning in the mid-1960s, and continuing to this day, teams of public health workers on horseback regularly visit the towns and ranches of the Beni, catching rodents. They kill a few and open them up to check the size of their spleens, since those that carry Machupo virus always have enlarged spleens. If the sentinel rodents have big spleens, officials know Machupo virus is around, and they institute a large-scale campaign of trapping and killing *Calomys callosus*.

This technique has worked well for Bolivian hemorrhagic fever because the virus causes a visible change in the animal that harbors it. For many rodent-borne or arthropod-borne viruses, though, this approach would be useless; the intermediate hosts of these viruses look and act perfectly normal even when they are infected. And for viruses with no vector at all, especially those that have no animal equivalent, there is no good way to anticipate the next outbreak short of waiting for the first few people to get sick. That is why some scientists, such as Edwin Kilbourne of Mt. Sinai, are worried that their peers are over-selling surveillance as a way to anticipate and prevent all infectious disease. As a member of Lederberg and Shope's committee that considered ways to anticipate the emergence of new microbes, Kilbourne concluded that watching for zoonotic (animal) infection—a surveillance approach that many of his colleagues favor—is of relatively little value. From a public health perspective, he says, not much can be done until the first wave of *human* infection occurs.

"I don't think we're going to prevent the next epidemic; that's an unreasonable expectation," says Kilbourne. "By the nature of things, I think things have to reach a threshold before they're perceived. That threshold may be very low: if you suddenly have three cases of rabies in a Vermont town, for example, everybody knows that's an epidemic and it quickly becomes a nationwide concern. But if you have three such cases in a small African village, it gets buried under the mass of other diseases there, so it might not be recognized or diagnosed." Listening posts in the developing world, such as those the World Health Organization sponsors to monitor trends in influenza, are of limited value, he says. "Such facilities have in the past uncovered a whole Pandora's box of viruses which we do not yet understand in terms of their relationship to human disease. Rockefeller Foundation expeditions back in the 1950s brought back all kinds of arboviruses, many of which are just sitting in deep freezes. So now what do you do? Do you make a vaccine against every one of these agents, when you don't even know what they do in people? In a sense, you have to wait until a human episode occurs."

A listening post, then, might be limited to doing just that—listening. Even if the outposts in the rain forest find a new virus in a chimpanzee or in a mosquito, virologists will not know the significance of that virus until something affects people, perhaps a continent away. "I think in a sense we have to be prepared to do what the Centers for Disease Control does so very well, and that is put out fires," Kilbourne says. "It's not intellectually very satisfying to wait to react to a situation, but I think there's only so much preliminary planning you can do. I think the preliminary planning has to focus on what you do when the emergency happens: Is your fire company well drilled? Are they ready to act, or are they sitting around the station house for months?"

No matter how highly refined the detection system, it works only if the pathogen itself offers some unwitting cooperation. Even the best-drilled response team depends on some lucky breaks if its members are to make sense out of the pattern of emerging diseases. To return for a moment to AIDS: detection of this new disease was something of a fluke, since its transmission pattern did not allow for the handy mechanisms afforded by SLE or Machupo virus. The human immunodeficiency virus does not infect animals, so a flock of sentinel chickens or even sentinel monkeys would have turned up nothing unusual. It is not carried by a vector, so there would have been no use in monitoring populations of mosquitoes or rodents for strange new viruses in their systems. But HIV did have one characteristic that made its detection possible: it caused symptoms so bizarre that even less-than-vigilant physicians could see that something new was going on. If AIDS patients had developed common infections instead of highly unusual ones; if the first population to be struck had been a broader cross section of the population instead of a readily distinguishable community concentrated in a few big cities; if the virus had remained in Africa, where the rapid death of previously healthy young people was comparatively common, instead of spreading to Europe and North America where such deaths raised suspicions—if any of these conditions had been different, the recognition of a new global pandemic might easily have taken years longer than it did.

Scientists do have a respectable track record monitoring one familiar virus, influenza, which travels along well-trodden paths that extend to every continent on earth. To anticipate new strains of influenza, scientists from the United States and Europe annually join forces with their counterparts in China. Until relatively recently, U.S. scientists from the CDC made annual pilgrimages to try to hunt down new influenza epidemics in person. It was an operation that depended largely on serendipity. If the Americans happened to be in China during a two- or three-week period when no new illnesses emerged, the plan was foiled; they had trouble during the remainder of the flu season keeping a step ahead of the virus, which is capable of traveling halfway around the world in a matter of months. But now

CDC has a new arrangement, begun in 1988, in which Chinese laboratories at several predetermined sites throughout the mainland are provided with American equipment to do surveillance on their own and report the results back to CDC headquarters in Atlanta. The Chinese labs now have eggs in which to grow the influenza virus; microscopes with which to examine the virus; and, in the central laboratory in Beijing, a freeze-dryer to preserve the viral isolates and ship them back to the American facility. The first year the program was in place, the CDC team in Atlanta received about 120 influenza isolates from all across China.

But even with this national and international organization, influenza continues to outfox us. Because it keeps eluding perfect detection, influenza helps scientists remain conscious of the notion of hubris— the belief that man can be felled by his own inflated self-image if he fails to take into account the limits of his ability to control the world. As Lederberg is fond of saying, in the battle over dominance of life-forms on the planet earth, the final struggle will be between man and microbe. And there is no assurance that human beings will emerge the winner.

"The concept of hubris, the idea that we know what's going on and we have control over it, is wrong," says Jonathan Mann, who now teaches at the Harvard School of Public Health. "As we speak, the next epidemic may actually be under way. I don't think we'll necessarily find it, but I do think that we must try." To try to anticipate the next AIDS, Mann has proposed a scheme just a bit different from D. A. Henderson's, a program he calls a "global pathogen watch." He suggests relying less on traditional public health approaches—field epidemiologists, laboratories, tests of sentinel animals—and more on interdisciplinary approaches embracing "psychologists, sociologists, and anthropologists, as well as virologists and specialists in communicable diseases, to develop creative ways of uncovering patterns of health and disease." Among his specific recommendations is the reliance on village elders to convey information about unusual illnesses. Mothers and grandmothers, he says, are often the first to notice when a strange new malady grips their families and friends. They and other local historians can often provide far more epidemiological information than can the best-trained American physicians, who might have a harder time separating out the old terrible diseases from the new terrible ones. "We must be prepared to look for evidence of a pathogen that is not yet known," he says. "Thus, we must think anew about the basic problem."

One possible benefit of global surveillance is perhaps not immediately apparent to those whose primary goal is the prevention of a new viral plague. But it might well be that another type of pathogen threat also will be uncovered by international monitoring: the threat of covert activities in biological warfare. Biological warfare, once considered a relic of the past—especially since it is prohibited by the Geneva Convention—made headlines in 1991 during the brief war in the Persian Gulf. American soldiers and Israeli civilians decked out in gas masks were an eerie reminder that even major military powers could easily be undone by a microscopic pathogen unleashed by a madman. Since 1972, when more than one hundred nations signed the Biological Weapons Convention, the development, production, and stockpiling of biological weapons have been banned. But there have been a few loopholes to this blanket proscription. Such weapons can be made and stored "for defensive purposes," and research into developing new *offensive* biological weapons is permitted to the extent necessary "to determine what defensive measures are required."

Modern molecular biology makes it difficult to differentiate offensive from defensive weapons research. If military scientists want to develop a vaccine against a virus thought to be under development for offensive strategic use, their first step would be to clone the virus itself and figure out how it works—a step that looks a lot like developing a biological weapon. In addition, the advent of genetic engineering allows scientists to manipulate pathogens in a way the signers of the 1972 treaty could never have anticipated. Now that molecular biologists can separate the genes of a virus according to function—those that cause the disease itself, and those that stimulate the body to mount an immune response—they have ways to deliver the toxicity without the immunogenicity. This could lead to biological weapons of unprecedented virulence.

There is no question that highly virulent strains of animal and human pathogens, tinkered with in the laboratory to permit dissemination through the air, would be a powerful military or terrorist weapon. Such a biological agent could destroy the economic and social underpinnings of even the most powerful of nations. What surveillance could do, in theory, is detect the occurrence of surprising animal and human diseases that could be traced back to biological weapons research being conducted in the broad loophole of this treaty. It would provide specific ammunition either to support or to refute the rumors that have circulated for years about one country or another's engaging in covert research, beginning with accusations coming from Cuba that the U.S. Central Intelligence Agency unleashed dangerous microbes on the island from 1971 through 1981.

"A program of global epidemiologic surveillance of major plant, animal, and human diseases is necessary to provide the capacity to detect (and thereby deter) clandestine biological attack," writes Mark Wheelis, a microbiologist at the University of California at Davis, who has spent much of his professional life thinking about how to prevent the development and use of germ warfare agents. A global surveillance system, according to Wheelis, "would make it very much easier for the world scientific community to deal with allegations of covert biological warfare, thereby deterring not only the activity itself but also reckless and destabilizing accusations."

Biological warfare is the more sinister side of a movement toward a new use of viruses that can also be quite beneficial: as agents doing humans' bidding rather than as messengers of doom. In laboratories around the world, viruses have become an important tool for procedures as varied as genetic engineering, vaccine production, and cancer prevention. In each case, a satisfying turnabout has occurred: now the virus is working for, rather than against, the best interests of human beings. Like a police informant who is wired for sound and sent back onto the streets to behave the way he always did, but this time with loftier goals, the virus has been lassoed and altered according to the whims of science. It is still allowed to infiltrate the cell, much as the informant is allowed to infiltrate the Mob or the drug ring, but it does so now at the behest of a new boss—us.

# 9

# *Viral Domestication*

The man was young, not yet out of his twenties, but he was clearly very sick. Pale, gaunt, with his blond hair thinning and askew, he could not keep his head up and was too weak to speak; he made no attempt to eat the hospital lunch in front of him. The man's mother was there, telling the doctor how out of sorts her son was that morning. "That's to be expected with the morphine we're giving you," said the doctor, W. French Anderson of the National Institutes of Health, addressing his patient directly. "But you know, there are thousands of people out there who would kill to get the very stuff you've been taking that makes you feel this way."

At this comment, the young man smiled broadly, nodding his wobbly head. "That's the first smile anyone's gotten out of him all morning," his mother said. Rotten as he felt, this young man knew he was lucky—as lucky as someone with malignant melanoma, a deadly form of skin cancer, could be. He was at the NIH Clinical Center in Bethesda, Maryland, to receive experimental anticancer drugs from the National Cancer Institute, known by the shorthand TIL (which stands for tumor-infiltrating lymphocytes) and IL-2 (interleukin-2). And he was awaiting the start of an even more dramatic step in cutting-edge cancer treatment, to be directed by Anderson—a treatment known as gene therapy. At that very moment (it was November 1990), the man's own TIL cells were being bathed in the lab in a culture mixed with two components: genes that produced a natural anticancer chemical called tumor necrosis factor, and a special conveyor molecule that would get those genes into the TIL cells for delivery to the melanoma itself. That conveyer molecule, essential for getting the genes into the TIL cells and, ultimately, directly to the young man's tumor, was a laboratory strain of a mouse retrovirus.

As lymphocytes with a particular affinity for cancer cells, TIL cells have long been thought of as a biological guided missile. They are the body's direct response to a cancer challenge; they exist only in cancer patients. But even though they tend to find their mark, in 60 or 70 percent of cases they are not by themselves powerful enough to vanquish the frenetically dividing cells of a tumor. Gene therapy is a way to give those TIL cells something extra to fight with once they hit their target. It upgrades the payload on these biological guided missiles from a few sticks of dynamite to something more on the order of a neutron bomb. The analogy is not a casual one, since tumor necrosis factor is so powerful that, when given by injection, it easily sacrifices the patient as part of its attack on the tumor. By hooking tumor necrosis factor onto TIL cells, which go directly to the cancer site, scientists hope to enhance the TIL cell's killing power where it is needed, and to spare the rest of the body.

As it turned out, the young man smiling at Anderson's feeble joke never got a chance at this new form of cancer therapy. Before the gene-boosted retrovirus had invaded his TIL cells in the culture dish, he died. But two other melanoma patients were luckier; their lab preparations were ready before the cancer overtook them. In January 1991, a twenty-nine-year-old woman and a forty-two-year-old man, under the direction of Anderson and his colleague Steven Rosenberg, became the world's first cancer patients to be treated with gene therapy. Just eleven other human beings had received gene-engineered cells before these two, ten for demonstration purposes only; their gene-boosted cells contained no therapeutic genes. The eleventh patient was a child being treated for an inherited childhood disorder quite unrelated to cancer, a disorder for which a second child would begin gene therapy treatment within days. One year later, the younger melanoma patient was still alive (the man died in early September), and the children— two girls, ages five and ten—were doing much better than their doctors had dared hope.

All four of these pioneering patients owed their improvement to that odd mouse retrovirus, which got the genes these people needed to the places where the genes were supposed to go. And because of this, it can be said that gene therapy represents another, more metaphoric, form of viral emergence. What is emerging here is a new view of the virus: not simply as the infernal competitor of mankind, not simply as the perennial microscopic foe, but as an agent that can be harnessed and domesticated to suit man's own ends. Gene therapy provides a new perspective on the relationship between viruses and people. In developing this treatment, medical science has taken the very quality that makes viruses so difficult to eradicate, and has twisted that quality to the benefit of human beings.

Even back in 1962, the eminent biologist René Dubos knew that microbes did not necessarily hold humanity in their sway. "Man makes use of microbial life for all sorts of practical ends," he wrote in his collection of essays, *The Unseen World*. Examples of microbial "domestication"—a word he coined to convey its similarity to the domestication of wild animals—were easy to enumerate: "Man makes a valuable pigment of the ochre produced by bacteria from iron; he cultivates in factories the yeast blown by the wind on to his grapes; he takes the very molds that spoil his foodstuffs, and with them he converts milk into savory cheese or he manufactures drugs to combat diseases; even more extraordinary, he takes the most virulent infectious agents and he finds ways to modify them or to use their products in order to develop vaccines for increasing resistance to infection."

Contemporary microbial domestication bears as little resemblance to the traditional approaches listed by Dubos—which, a mere thirty years later, seem almost quaint—as Gregor Mendel's work with sweet peas in the monastery garden bears to the DNA sequencing taking place in a modern molecular biology lab. Domestication today involves a rearrangement of the very structure of the virus, turning it into a microscopic Trojan horse that gets genes or other substances right into the core of the cell. That harnessed virus is called a *vector*, from the Latin word "to carry." The trick in domesticating such a vector is to maintain the virus's ability to get inside a cell, while at the same time stripping it of its ability to cause disease. This means eliminating all viral genes controlling replication, which determine virulence. The protein coat remains, and with it the virus's methods of camouflage or whatever other subterfuge it uses to penetrate the cell's protective membrane and insinuate itself into the cell's nucleus. In place of the replication genes, the viral vector carries other, totally unrelated genes—not even viral genes at all—that are the payload of gene therapy.

Before 1983, when the first retroviral vector was developed, molecular geneticists tried to get genes past the cell's protective membrane by altering the membrane with chemicals, electricity, even brute force. They used calcium phosphate, for instance, to punch holes in the cell membrane, making it easier for a gene to find its way into the interior. But once past the membrane, genes had a tendency to stay in the cellular material, or cytoplasm, and to go no farther. Where they were really needed, of course, was in the nucleus, alongside the cell's own genes. In the early days of gene therapy, about one cell in a million actually carried the gene into the nucleus and showed by its functioning that the gene was working. That wasn't good enough. Neither was another method tried in those days: the literal squirting of a gene directly into the nucleus, a process known as micro-injection. Only a few cell types, most notably muscle cells, responded well to micro-injection. And even they had to be micro-injected one at a time—an enormous obstacle for treatments like those used today, which require up to thirty *billion* gene-boosted blood cells for every infusion.

The slow progress of gene therapy research shifted gears in the early 1980s, when Richard Mulligan had an epiphany. Mulligan, a molecular biologist at the Whitehead Institute for Biomedical Research at the Massachusetts Institute of Technology, reasoned that one way to get genes into a cell nucleus would be by co-opting nature's own best invader, the virus. The virus he first used was a monkey retrovirus; scientists at NIH now depend on a mouse retrovirus known as Moloney murine leukemia virus. The retrovirus was the "obvious" choice for a gene vector, Mulligan says in retrospect: "You couldn't possibly have conceived of a better life cycle for effecting gene transfer." It can penetrate the cell nucleus and insert retroviral DNA into the cell's chromosomes, a process known as transduction; it is smaller than most other viruses, which means there are fewer infective genes to disarm; and, unlike most other viruses, it usually does not kill the cells it infects.

Once the vector was found, the next choice in the development of gene therapy was to send the vector to the right target cells. The first human cells to be transduced in vitro were those most accessible to lab manipulation: blood cells, bone marrow cells, and skin fibroblasts. If the genes could be put back at least into the cells where their product was most needed—since not every gene is turned on, or expressed, in every cell—gene therapy would have the greatest chance of a cure.

Bone marrow cells originally seemed like the ideal cells for transduction. The cells are accessible, even though to obtain them meant
inserting a long hollow needle into the patient's hip under general an-
esthesia. Even more important, a small proportion of bone marrow
cells, called stem cells, is the granddaddy of all blood cells: they give
rise not only to lymphocytes but to all the other cells of the circulatory
system, such as red blood cells, plasma cells, and macrophages. Unlike
lymphocytes, which have a life span of a few months, stem cells last
forever. It was thought that insertion of a needed gene into a stem cell
would guarantee that all cells derived from that stem cell, for the rest
of the patient's life, would have the new gene too.

But stem cells turned out to be elusive targets. During the mid-
1980s, stem cells proved to be all but impossible to find, much less to
transduce. Fewer than one bone marrow cell in ten thousand is a stem
cell, and there's no good way to separate stem cells from the others.
In addition, transduction occurs only in cells that are dividing;
stem cells rarely divide. The odds against a retroviral vector's actually
getting through to a stem cell seemed to be enormous. The compro-
mise was to treat lymphocytes instead. Lymphocytes were easily accessible, and for the conditions first treated experimentally, they were
appropriate targets. The only problem with treating lymphocytes
rather than their precursors, stem cells, is that gene therapy done
this way would require frequent reinfusion of cells, so that new
generations of genetically engineered lymphocytes could be continually introduced.

To begin therapy with gene-boosted lymphocytes, the scientist
takes the patient's blood in a procedure much like an ordinary blood
donation, and uses a ficoll polymer gradient—a lab technique that sep-
arates solutions into their component parts on the basis of density—
to separate the lymphocytes from all other blood cells. (The lymphocytes line up along a layer from which they can easily be extracted.)
Back in the laboratory, the scientist grows the lymphocytes in huge
batches, using a special culture medium that encourages their growth,
and finally adds a solution containing the retroviral vector into which
the relevant gene has been spliced. Within hours, the gene-engineered
virus penetrates at least some of the lymphocytes.

The first patient treated in this way, in September 1990, was a little
girl with an extremely rare genetic condition, affecting just fifteen or
twenty children worldwide, called ADA deficiency. When a particular

gene on chromosome 20 is missing, the body cannot manufacture ADA, an enzyme required to prevent toxins from accumulating in the budding immune system. Ultimately, the immune system never forms, and children with ADA deficiency develop a disorder known as severe combined immune deficiency (SCID)—which means they are as vulnerable to infection as are people with AIDS. In the 1970s, a young boy named David who had SCID received much publicity for his lifetime spent in a sterile enclosure, leading to his nickname "the bubble boy," to keep him from dying as a result of exposure to a pathogen his body could not fight off. Children with ADA deficiency no longer have to live in isolation the way young David did. The lucky ones, who have siblings with perfectly matched tissue, can receive bone marrow transplants and essentially be cured. The less lucky ones can gamble on a bone marrow transplant from a near match, which works only about half the time. That is what David did in 1984, but his gamble proved to be a fatal one—though not for the reason his doctors had expected. David received a bone marrow donation from his sister, and his body never rejected her cells. But in those cells, it turned out, were some latent Epstein-Barr viruses that had been un-detected, probably remnants of his sister's earlier bout with mono-nucleosis. Those viruses were perfectly harmless to David's sister—whose healthy immune system kept them in check—but proved to be overwhelming to him. When the boy died four months after the transplant at the age of twelve, his body was found to be riddled with bean-size tumors, every one of them containing the Epstein-Barr virus.

Children with ADA deficiency now have an alternative to risky bone marrow transplants: they can take a drug called PEG-ADA. PEG-ADA delivers ADA directly into the bloodstream while pro-tecting the enzyme from degradation by the body. But even with weekly injections, children with ADA deficiency still have a lympho-cyte count that is about one-half of normal, and for them, schoolrooms and other places where children congregate can mean a high risk of infection. The medication is expensive as well; a year's supply of PEG-ADA, according to the manufacturer, costs about $300,000; a year's worth of gene therapy treatments probably costs about $100,000, but during these early experiments the NIH absorbs the expenses. When the limitations of PEG-ADA therapy are taken into account, the fam-ilies of the two young girls being treated by Anderson and his NIH colleagues, Michael Blaese and Kenneth Culver, decided that gene therapy, uncertain as it was, offered the best chance for their children to live relatively normal lives.

Gene therapy does not require that a new gene go directly to the right spot on the chromosome where it ordinarily resides. Instead, all that is needed is for the hybrid vector (the crippled virus plus the genes that have been inserted into it) to do what viruses do best: insert itself into the genetic material of the cell and commandeer the cell's reproductive machinery into making more copies of the virus product. Then, if the engineered virus contains a gene responsible for the secretion of a particular protein, its presence in the nucleus of the target cell—no matter where along the genome it lodges—should be enough to get that protein manufacture going.

But the lack of control over precisely where a gene goes, even when it is packed inside a viral vector, is one of the great shortcomings of gene therapy. There is always the chance that the gene will end up somewhere where it could do damage. It could, for instance, settle in next to a cancer-causing gene and somehow turn it on and lead to cancer. That is why investigators are now studying different viruses that should help them aim the genes more directly.

One of the most promising of such new viral vector systems involves adenovirus. The adenovirus ordinarily infects the lung epithelial cells, those lining the walls of the lung's tiny air sacs, known as bronchioles. It causes the symptoms of the common cold (which is more often caused by another family of viruses, the rhinoviruses), sore throat, croup, pneumonia, bronchitis, and other upper respiratory infections. Some cases of "the flu" are not caused by influenza virus at all, but by adenovirus. Because of its tropism for lung cells, adenovirus has been considered for the treatment of some common genetic conditions affecting the lungs, most prominently cystic fibrosis. Recently, scientists at the National Heart, Lung, and Blood Institute at NIH conducted a series of experiments involving gene therapy for lung diseases that used cultured lung epithelial cells transduced with gene-altered adenoviruses. "In vitro, in terms of expression [getting the transduced cells to manufacture the gene product of interest], adenovirus is far superior to retroviruses," says Ronald Crystal of the NIH, who conducted the work in collaboration with scientists at the Institut

Gustave-Roussy and the biotech company Transgene, both in France. "This is a very powerful vector."

Crystal's team transduced adenoviruses with the gene for alpha 1-antitrypsin, the absence of which is responsible for a relatively rare form of hereditary emphysema. Alpha 1-antitrypsin is a chemical that normally blocks the action of elastase, an enyzme that digests protein. When alpha 1-antitrypsin is missing, elastase continues its digestive actions unabated, with the result that the linings of the lungs' air sacs are eventually eaten away and the patient is unable to breathe. Because no gene therapy alternative exists as yet, people with this inherited defect in the alpha 1-antitrypsin gene are now treated with weekly injections of the enzyme itself, much as people with ADA deficiency—pending the outcome of the NIH trials—are treated with weekly shots of PEG-ADA.

But gene therapy for emphysema is appropriate only for a very rare emphysema, the genetic form, which affects some 2 percent of all emphysema patients. A more common use of the adenovirus vector, once it is developed for humans, would be in the treatment of cystic fibrosis, the most common genetic defect in North America. Cystic fibrosis, which primarily affects Caucasians of European descent, causes an accumulation of abnormally thick mucus in the lungs, leading to frequent infections and scarring; most patients die during adolescence or young adulthood. The gene for cystic fibrosis has already been identified in human patients and manufactured in the lab. If it can be inserted into an adenovirus vector and the vector then converted into an aerosol form, gene therapy for cystic fibrosis could be as convenient as is inhalant treatment for asthma.

For people now struggling with the imperfect therapies available for this terrible genetic disease, the prospect of thousands of cystic fibrosis patients' carrying around portable spritzers filled with genetically altered viruses may seem like a blessed relief. But it also raises fears about environmental hazards to other individuals. "If the virus can be aerosolized to get into the lungs, it can also get out," says Barrie Carter, chair of the biosafety committee of the National Institutes of Health. "And what do you do when an altered adenovirus gets out?" This would be different from the accidental release of an altered mouse retrovirus (the only viral vector used to date in human gene therapy), which is incapable of infecting humans even when it is fully functional. This would be a human adenovirus, which causes human diseases, that was genetically altered to make it spread more efficiently through the air. Additional genetic changes would be built in to make it less infectious to people once it is aerosolized, but even with those changes, this adenovirus vector could pose health risks quite different from those posed by gene therapy involving the Moloney murine leukemia virus.

Other types of viral vectors also are being developed by university scientists and biotechnology companies around the world. The holy grail of molecular geneticists is an injectable vector—a viral solution that a doctor or nurse can pull off a shelf and administer as easily as a flu vaccine. It would make human gene therapy available to anyone, anyplace—not just to individuals with access to major medical centers, the only sites currently capable of performing the elaborate blood-drawing procedures necessary to remove and infuse lymphocytes. Once an injectable vector is developed, the potential use of gene therapy can go beyond the treatment of genetic diseases to the treatment of some of the industrialized world's leading killers, among them:

- Cancer. Scientists are planning to use viral vectors to deliver a "suicide gene" directly to cancer cells, to make them more vulnerable to destruction by chemotherapeutic drugs. If a gene for drug sensitivity can be inserted into tumor cells, then exposure to that drug may cause the gene-engineered cells selectively to self-destruct.
- Cardiovascular disease. Current surgical treatment of blocked blood vessels sometimes involves insertion of a tiny stent, a piece of metal that physically holds the vessel open. An alternative treatment is bypass surgery, in which a new vessel, known as a vascular graft, is inserted to replace a clogged one. But in a significant number of both of these implants, blood clots rapidly develop. Scientists are trying to coat both types of implants with cells containing the gene for tissue plasminogen activator, a natural clot dissolver, to avoid this complication of surgery.
- High cholesterol. Molecular geneticists are trying to develop "neo-organs," structures made of artificial material that can receive a desired new gene. Among the most promising are neo-organs placed near the liver, into which is inserted a viral vector containing a gene that attracts cholesterol. Connected to the liver by a network of new blood vessels, this neo-organ could pull

excess cholesterol from the bloodstream—where it does its cardiovascular damage—and into the liver for excretion.

- AIDS. Scientists at the National Institutes of Health are experimenting with the treatment of AIDS by delivering the gene for
soluble $CD_4$, a protein that lies on the surface of the T cells
vulnerable to infection by HIV, which has been fused with another protein. The attachment of HIV to the $CD_4$ protein is the
mechanism by which it gains access to the lymphocytes. If
enough mock $CD_4$ can be kept circulating in the bloodstream,
these decoys may snare the AIDS virus and keep it from getting
to the immune system cells at all.

A variation on traditional gene therapy is known as intracellular immunization, a kind of general antivirus vaccine that is also genetically
engineered. As with gene therapy for cancer or ADA deficiency, the
patient's blood cells are mixed with a virus vector that contains a specific gene. In this case, though, the vector carries a gene known to
inhibit the growth of the virus of interest. It is not unlike the release
into the wild of mutant mosquitoes as an attempt to control arboviruses. With arboviruses, the released mosquitoes were genetically engineered to make them incapable of carrying viruses. Their release in
sufficient numbers meant they would outpace vector-competent mosquitoes in the competition for mating and propagation. Similarly, if
enough mutant viruses are delivered to a cell—viruses that were genetically manipulated so they lack a particular gene needed to replicate—they would in theory edge out other infectious viruses in *their*
struggle for reproduction, the competition for the same scarce cellular
resources. The crippled, harmless viruses would dominate, and the
infectious viruses would die out.

Intracellular immunization was first used in 1988 by scientists at
the Carnegie Institution of Washington in Baltimore, who worked
with herpes simplex virus and mouse cells in culture. They primed
the mouse cells with a protein known as viral protein 16 (VP-16), a
component of the herpesvirus that allows it to begin replication in an
infected cell. VP-16 is like the virus's On switch. But the trick in this
experiment was that the VP-16 in the cultured cells was a mutant.
Using a virus vector, the scientists bathed the mouse cells in VP-16
protein that was missing some essential genes, the loss of which ren-

dered it incapable of turning on the herpesvirus. When a functional herpesvirus was then introduced into the cell culture, its own VP-16 could never get going; the mutant VP-16 was present in such huge numbers that it beat out the virus's own functional VP-16 for binding sites on the host cell's proteins. Result: the replication of the herpesvirus could never get started, so the cells never became infected.

The most dramatic potential application of intracellular immunization is as a deterrent to infection with HIV. The AIDS virus also has an activator gene, called *tat*, which initiates the replication of HIV. Now that scientists know which region of the *tat* gene is critical to its functioning, they can begin the same experiment that was successful using herpes simplex's VP-16: they can deliver a crippled *tat* gene, via a viral vector, directly to an animal's lymphocytes or bone marrow cells and see whether the mutant gene predominates and shuts out the functioning of HIV's own *tat*.

Ali Maow Maalin's brush with death was historic. Maalin was a hospital cook living in the small town of Merka in Somalia, along Africa's eastern coast. On October 13, 1977, a car stopped at the hospital where Maalin worked, and the driver asked directions to the smallpox isolation camp. Maalin hopped in to guide him to the director of the local smallpox surveillance team, who lived less than a kilometer away. In the car were two children bound for the isolation camp: a six-year-old girl with a severe case of smallpox, and a boy of two whose infection was less advanced. Somalia was at war with Ethiopia, the only remaining country where smallpox was endemic, and thousands of smallpox-infected refugees had been crossing the Ogaden desert into Somalia since the previous fall. In the space of a little over a year, a total of 3,229 cases of smallpox were recorded in Somalia.

As far as the story of Ali Maow Maalin is concerned, the friendly gesture of riding in a car with two sick children was all it took to guarantee him a place in the medical books. Because Maalin's, it turned out, was the last case of smallpox on earth.

Maalin left work early on October 22, nine days after his fateful car ride. On October 25, he was admitted to the hospital and treated for malaria. The next day a rash developed, and the day after that he was sent home with a diagnosis of chickenpox. Maalin suspected that what he really had was smallpox, but he kept silent—he was afraid of the quarantine and isolation procedures he would face. Soon, though, the truth was hard to avoid. His rash turned pustular by the end of the month, with large weeping sores, especially on his arms and legs all the way to the palms of his hands and the soles of his feet. On October 30, a nurse in Maalin's hospital contacted the local health authorities to report a case of smallpox. Maalin was forced to remain isolated within his house; a twenty-four-hour guard was placed on watch. Two days later, he was moved to the smallpox isolation camp. His fever and rash worsened, but he eventually recovered. At the end of November he was sent home, cured. "During his illness, Ali had exposed 161 persons, 91 of them face to face," recalls Abram Benenson, a professor at the Graduate School of Public Health at San Diego State University. "None of these contacts developed smallpox. No subsequent cases of naturally acquired smallpox have been discovered on the face of the earth."

It took a long time to get to that last human case of smallpox. The virus is probably the longest-lived human pathogen in history, with documentation of smallpox dating back at least two thousand years. Indeed, some medical historians believe that Ramses V, the Egyptian pharaoh who died in 1157 B.C., showed evidence of pockmarks on his mummified face. It seems poetic justice that smallpox, one of our oldest scourges, is also the first virus that has been totally eliminated because of the actions of human beings. And it seems even greater vindication to take the virus responsible for that elimination—the vaccinia virus—and domesticate it sufficiently to help eradicate other viral diseases.

All biology students are taught the story of Edward Jenner and the very first vaccination in medical history. Jenner, as legend has it, was a general physician in the English region of Gloucestershire who noticed that milkmaids never seemed to get smallpox. This accounted for their ability to keep their fresh peaches-and-cream complexions far into adulthood; their faces never showed signs of the pockmarks that scarred smallpox survivors. From this observation, Jenner theorized that the cowpox agent they were exposed to while they milked their dairy cows—which caused a milder disease, with pockmarks generally restricted to the hands and arms—was somehow protecting them. Jenner followed up his hunch in 1789, by inoculating his infant son, Edward junior, with material gathered from sores of a related zoonosis, swinepox. Then Jenner injected the baby with live smallpox, and little

Edward did not get sick. In 1798, an epizootic of cowpox raged through Gloucestershire, and for the first time Jenner had enough material to put his hypothesis to a meaningful test. This was years before viruses were even imagined, of course, but Jenner presumed that something in the exudate from the sores of pox-infected cattle was involved in the disease process. So he found eight young children willing to serve as experimental subjects, scratched some cowpox exudate into his subjects' skin—and waited. Each one (except one boy, who died of other causes) developed a single pockmark at the site of the scratch, which eventually crusted over and fell away. Each one resisted infection after subsequent inoculation with smallpox.

We immortalize Jenner's contribution by the very name given to vaccinations, which comes from the Latin word *vacca*, for "cow." But what we fail to memorialize is the fact that the virus used in most modern smallpox preparations, the vaccinia virus, is not quite the cowpox virus. Somehow, for reasons no one really understands, a new virus emerged in the smallpox vaccine preparation early in the twentieth century. That new virus was given a name with the same Latin root as cowpox, but it was a different virus. It was called the vaccinia virus.

Despite the common name, though, vaccinia is quite different from cowpox. As Frank Fenner, the Australian virologist and international authority on poxviruses, has written, vaccinia's origins are a complete mystery. "Analysis of the genetic material of several strains of smallpox, vaccinia, and cowpox viruses has confirmed [that] the structure of their genomes is such that the 'transformation' of one species into another, a hypothesis previously advanced for the origin of vaccinia virus, is impossible." Vaccinia virus is the only poxvirus with no natural reservoir. The only time it ever infects an animal species—and it has in the past been shown to infect cows, camels, buffalo, and pigs—is through exposure to a recently inoculated human being. Its only natural habitat today is in the research laboratory. But mysterious as its origins are, vaccinia's genetic structure is an open book. And because the virus is so carefully studied, and its track record as a vaccine is so long, many scientists consider it to be the most likely contender for domestication into the world's first genetically engineered vaccine.

The process of using a virus to prevent viral disease is, in essence, much the same process as that used in gene therapy. It involves calling into action a crippled virus to serve as a vector for getting particular genes into particular cells. One difference, though, is that the genes involved in recombinant vaccines are viral rather than human in origin. The purpose of those viral genes is to make the cells behave in a way that will stimulate an immune response. The genes, though carried in the vaccinia vector, are quite unrelated to vaccinia, just as train passengers have no relation whatsoever to the train itself.

Vaccinia virus is being used in most of these experiments because it is so big, so hardy, and so well understood. Vaccinia virus has proved itself to be the most successful vaccine in medical history. It is capable of living in the air for many days in a freeze-dried state without losing its infectivity, an important consideration if vaccines are to get to countries with poor refrigeration or other storage problems. In addition, vaccinia is quite large, for a virus. Like other poxviruses, its size approaches that of a small bacterium, making it part of the only group of viruses that can be seen under a normal light microscope. Vaccinia virus has a genome of some 185,000 nucleotides. When those nucleotides are stripped away, as part of the process of rendering a viral vector noninfectious, that leaves behind a lot of empty slots to fill. Genes coding for several viral proteins, from several different viruses, can be inserted into one of the many spaces available in the vaccinia genome. Vaccinia thus becomes a kind of "vaccine bus," carrying genes from as many as ten or twelve different viruses in a single injection. This would greatly simplify the goal of universal immunization; the fewer shots (and clinic visits) necessary for total protection, the more likely it is that each child will receive the full range of possible immunizations.

The first recombinant vaccinia vaccine is already being field-tested, but it is being targeted at animals. In 1990, a rabies vaccine using a vaccinia vector was released into the forests of Europe for the first time to immunize wild foxes, the major rabies carrier on the Continent. Obviously, scientists cannot bring a succession of wild animals into the veterinary clinic for rabies shots; something more creative must be done when the disease begins to overrun a community of free-living animals. (Epizootics of rabies among wild animals can translate into rabies infection of domestic animals—during a fight between a dog and a squirrel, for instance—and occasionally of people as well, either through the domestic animal or from direct exposure to the rabid wild animal in backyards, parks, or forests.) The method of delivery of the genetically engineered vaccinia vaccine works this way: animal food is

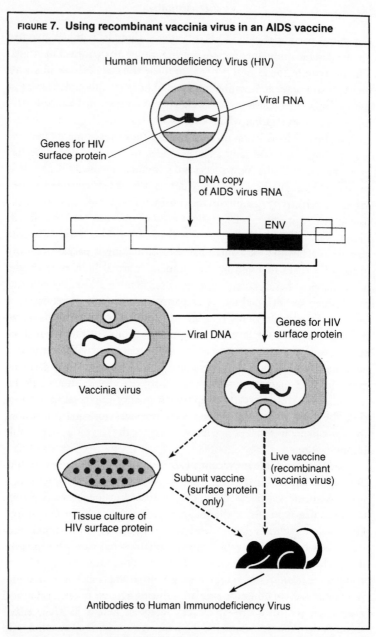

FIGURE 7. Using recombinant vaccinia virus in an AIDS vaccine

Human Immunodeficiency Virus (HIV)

Viral RNA

Genes for HIV
surface protein

DNA copy
of AIDS virus RNA

ENV

Viral DNA

Genes for HIV
surface protein

Vaccinia virus

Live vaccine
(recombinant
vaccinia virus)

Subunit vaccine
(surface protein
only)

Tissue culture of
HIV surface protein

Antibodies to Human Immunodeficiency Virus

Figure 7. In a new approach to vaccine development, scientists have tried to insert genes from the AIDS virus (ENV), which codes for the virus's surface proteins, into a vector derived from the vaccinia virus. This recombinant vaccine could then either be given directly to an experimental animal as a live vaccine, or grown in tissue culture to yield a subunit vaccine that delivers only the surface protein itself. In either case, the result should be the production of HIV antibodies in the animal.

delivered by air across rural areas of France and Belgium, where fox rabies is a major problem that regularly infects cattle from bites. The food has been mixed with genetically cripped vaccinia viruses into which antigens from the rabies virus have been spliced. When the animal eats the virus-laced food, it begins producing antibodies to the rabies antigen. Preliminary reports from Belgium in late 1991 indicated that of seventy-nine foxes subsequently caught and studied, all but one had antibodies sufficient to provide immunity to infection should the fox subsequently encounter the rabies virus itself.

Vaccinia is also being used as the basis for some human vaccines. Joel Dalrymple at the U.S. Army Medical Research Institute of Infectious Diseases, for instance, is involved in an effort to add genes from the Hantaan virus to a vaccinia vector. He considers the generations of experience with vaccinia to be a big plus in working with the virus, but adds an even more compelling reason: vaccinia replicates exclusively in the cytoplasm, rather than traveling into the cell nucleus. "Now, if you're going to make a vaccine and you're going to worry about foreign genes and their incorporation into the basic genetic material," Dalrymple says, "you would like something that stays in the cytoplasm. Vectors like retroviruses or tumor viruses become integrated into the chromosome of the cell," and the long-term hazards of getting a new gene into the host cell's genome are unknown. When vaccinia is a vector, Dalrymple says, "the nucleus is sacred; it's never violated. The human material in the nucleus never comes in contact with the virus replication in the cytoplasm."

Despite these benefits, though, vaccinia presents many problems in terms of its suitability as a vaccine vector, not the least of which is the difficulty in completely incapacitating the virus itself. No one born since 1977 (and even earlier in industrialized nations) has had any exposure to vaccinia through smallpox inoculation, so they are vulnerable to vaccinia infection now. If they receive a recombinant-vaccinia vaccine in which the vaccinia virus has been inadequately attenuated, they can get a lethal infection.

Vaccinia virus infection is especially dangerous for young children or adults with immune system abnormalities. During the smallpox eradication effort in Africa, postvaccine encephalitis occurred more often after vaccinia vaccination than after vaccination with other agents. This was a severe illness, with a death rate approaching 50 percent and a high rate of permanent brain damage among survivors.

Other less grave complications included eczema vaccinatum (a skin eruption that was almost as bad as smallpox), generalized vaccinia, exanthematous (rash) reactions, accidental infection of close contacts, and congenital vaccinia in children born to mothers who were vaccinated during pregnancy. These complications were worth enduring for the sake of protection from a deadly infection. But now that smallpox has been eradicated, the risk-benefit equation changes. If there is no need to expose any human being to the vaccinia virus for immunization purposes, why run the additional risk entailed in using vaccinia for a vector? It would be worth the risk only if the recombinant vaccine being developed were for a disease as deadly as smallpox—possibly, for example, for AIDS.

But a recombinant vaccinia vaccine for AIDS presents problems of its own. In 1991, some AIDS patients in Paris were given an experimental vaccine based on an immobilized strain of vaccinia, into which genes coding for HIV coat proteins were inserted. Although the investigators thought the vaccinia was attenuated, it apparently retained some residual infectivity—at least enough to overwhelm the already compromised immune systems of AIDS patients. Two of the subjects contracted a gangrenous vaccinia infection; both died.

Many of the experimental vaccines might get around the problem of an infectious vaccinia vaccine because of the very method used for inserting new genes into the vector. Whatever genes of interest are added—such as, in the case of the military effort, the Hantaan virus genes—they are inserted in a region of the vaccinia genome called the thymidine kinase (TK) gene, replacing the TK gene completely. The TK gene also happens to determine the virulence of vaccinia. Its removal to put in different genes, therefore, has the coincidental effect of further immobilizing the vaccinia in the vaccine preparation.

And many of the vaccinia vaccine experiments now under way don't involve the actual introduction of vaccinia into the subject at all; vaccinia is used only as a way to get a gene where it should go, and then its role is ended. At the National Institute of Allergy and Infectious Diseases, for instance, Bernard Moss is using vaccinia virus to get particular proteins from HIV into mammalian cells—a set of proteins that will enable those cells to act as mini-factories for an entirely new type of vaccine. Moss and his colleagues co-infected monkey cells with two vaccinia vectors, the first carrying a gene for RNA polymerase, which helps transcribe DNA into RNA; the second carrying a "switch" attached to the gene for gp160, the surface protein of HIV. Once infected, the monkey cell began producing RNA polymerase, which in turn bound to the switch and initiated production of gp160. In this way, Moss made an entirely synthetic HIV surface protein, which he could then use to inoculate volunteers at high risk of exposure to the virus. The benefit of this system is that, because the artificial protein is manufactured inside a mammalian cell, the three-dimensional shape of the gp160 is virtually identical to its native shape. This is probably important in the stimulation of a strong immune response. But this technique, while it relies on vaccinia to get started, would not involve actual introduction of the virus into human beings.

Recombinant vaccinia virus also forms the basis of a two-pronged pest eradication campaign just getting under way in Australia, in which vaccinia will be used to sterilize Australian foxes, while another familiar virus, myxomatosis, will be part of a complementary effort to sterilize Australian rabbits. This story of recombinant myxomatosis, which is still unfolding, is in many ways a satisfying conclusion to our exploration of emerging viruses. It brings us back to where we started, but with a twist.

Those imported European rabbits and their cousins, European red foxes, continue to plague Australia even forty years after Frank Fenner first tried the eradication-through-viruses approach there. Rabbits are still a nuisance, still destroying crops and forests, while predatory foxes have been blamed for the extinction of twenty species of marsupials and currently threaten endangered marsupials like numbats, woylies, and wallabies. So Australian scientists are becoming high-tech in their attack. Beginning in 1995, they hope to insert into live, infectious myxoma virus a rabbit sperm protein that will induce an immune response in female rabbits, thereby causing them to reject sperm and making them sterile. For foxes, scientists will do the same thing in recombinant vaccinia virus (which they're forced to use because no species-specific virus exists for foxes comparable to the specificity of myxoma virus for rabbits).

This sterilization-by-virus approach has its share of critics; even the scientists involved aren't sure yet just how safe it is. As Mark Bradley, a reproductive immunologist at Australia's Cooperative Research Centre for Biological Control of Vertebrate Pest Populations, told a reporter for *Science* magazine, "No country has ever tried to manage a pest species on this scale or in this way before. It raises questions across disciplines, from virology to immunology to the animals' social behavior and ecology. And we're trying to answer them all."

The National Institutes of Health provides spare accommodations even for its senior scientists. After more than a quarter-century on the staff of the National Heart, Lung, and Blood Institute, after being generally acknowledged as the world's leading authority and most energetic proponent of human gene therapy, W. French Anderson had quarters consisting of one long, narrow office and one long, almost as narrow meeting room. (He left in late 1992 for perhaps more spacious surroundings at the University of Southern California.) Into that meeting room were crammed the photocopier, fax machine, computer, coffee maker, journal collection, and conference table—and, during the twice-weekly laboratory meetings, fifteen or twenty enthusiastic scientists. Just above the Xerox machine was a framed quotation of some lines from *Hamlet*, lines that offer a glimpse into how human gene therapy was being perceived by the men and women who devote their professional lives to seeing it come to fruition:

> *Diseases desperate grown*
> *By desperate appliance are relieved*
> *Or not at all.*

The diseases we now confront are indeed desperate: cancer, genetic defects, AIDS. It remains to be seen how "desperate" are the methods developed to treat those diseases, and just how many of them can be truly "relieved" by genetically based approaches. But one thing is clear: medical science would be helpless to loosen the grip of these diseases on mankind without its new ability to harness a few viruses to serve as mediators. And because viral domestication is the first step in genetic engineering, the technique itself is testimony to the fact that everything in the universe—even the lowly, parasitic virus—has buried within it the ability to help the very creatures that it also can undo.

# 10

# *Toward a New Biology*

In December 1989, the head of an ad hoc government team had grim news for the leading American experts in tropical medicine. Refugees from a civil war in Changa, a small country in sub-Saharan Africa, were fleeing by the thousands into neighboring Basangani. There, in squalid refugee camps, they were dying of a bizarre illness that looked like a highly infectious form of the dreaded African virus called Ebola fever. Relief workers were dying too. "Three hundred refugees and four relief workers have died in the past month," said the government official, Llewellyn Legters, chairman of preventive medicine and biometrics at the Uniformed Services University of the Health Sciences in Bethesda, Maryland. "Many volunteer workers have given up and returned to Europe and the Americas." But many of those who came home were already infected with the air-borne virus, and they were unwittingly spreading it all over the globe.

The news was serious indeed. But it was faked. It was part of a "medical war game" staged by Legters at the annual meeting of the American Society of Tropical Medicine and Hygiene, a sort of dry run for a global virus emergency. And after one long day of this deadly simulation, Legters had made his point: the United States and Europe are not prepared for a lethal new virus sneaking across their national borders.

As the scenario got more and more gory, it became clear that the Centers for Disease Control had too few resources to handle a budding emergency on two continents at once. The CDC had a total of just four experts in hemorrhagic fevers, too few to dispatch a team to Basangani and U.S. military bases at the same time. Similarly, the agency's single portable high-containment research laboratory had to be located either in Africa or somewhere in the United States, and no one knew which was the better place. Nor did anyone know where to send the army's only isolation field hospital unit. Should it go to a hospital in the United States, or to one of the three overcrowded refugee camps in Basangani?

Complicating the story were the sick American volunteers. One week into the script, several relief workers had already died of this mysterious fever: two in a private hospital in Washington, D.C., one in Africa, two in Basangani. They had doubtless exposed their American contacts, health care workers, and even morticians to the bizarre virus. And as more relief workers fled the area, they were each exposing hundreds more people to the virus on the twenty-one-hour airplane flight back home.

The simulation dramatized the importance of inter-disciplinary coordination in dealing with a viral emergency. The "ad hoc government team"—really eight hundred tropical medicine experts who were attending the conference that year—kept asking for advice from many experts: from clinicians caring for patients in the United States and abroad, from epidemiologists tracking down the source of the infection, from molecular biologists examining African blood samples for signs of a new virus, from military strategists and diplomats monitoring infection among troops sent to Changa as part of an international peacekeeping force. Letgers had made his point: no progress would be made without the collaboration of physicians, epidemiologists, laboratory scientists, policymakers. Each one was just as important as any other.

But getting these experts to talk to one another is no easy matter. They tend to speak entirely different languages and to view their object of study, the virus, from entirely different vantage points. The language gap is huge. Like most other lay people, I had been quite unaware of the little fiefdoms into which the scientific enterprise is usually divided; I never anticipated that, even within the general discipline of biology, scientists consider themselves to be defined more by their differences than by their similarities. As an outsider, I always had assumed that scientists were all playing by the same rules, with the same goals. I was wrong.

The late DeWitt ("Hans") Stetten, for many years the scientific director of the National Institutes of Health and a widely respected thinker on the culture of science, had a handy way of illustrating the rifts dividing the different scientific subspecialties. He once drew a line of all the disciplines of science and near-science, written out left to right in the following order: mathematics, physics, chemistry, biology, medicine, economics, psychology, political science. (He didn't include ecology or environmental science, but that would probably be farthest to the right on this continuum.) If $X$ marks the spot of a scientist's own discipline on this line, he or she typically would consider everything to the left of that $X$ "hard science," and everything to the right, "soft." Calling something soft science, of course, is the most biting criticism a scientist can wield; it means lacking rigor, weak on data, impressionistic. An attitude like this among biologists—who themselves are not so far left on Stetten's line—automatically rules out a large group of disciplines that could enlighten their own investigations.

But if scientists continue to stick to their own tiny niches, the quality of the science itself will suffer. That is perhaps the most compelling reason to break down disciplinary barriers. And that is one of the most promising possibilities to come out of the study of emerging viruses. Such a coordinated study can lead to an emergence of its own: the emergence of a new biology. There is no useful way to study viral emergence without incorporating the perspectives of many disciplines, both from within the field of biology and from other, apparently unrelated, fields as well. This requires the amalgamation of many branches of biology—molecular genetics, cell biology, field biology, entomology, ethology, population biology—as well as many branches of clinical medicine, such as infectious disease, immunology, and epidemiology. Scientists also will need to rely on the expertise of people they have little opportunity to meet or converse with even inside their own universities or research centers: chemists, anthropologists, geographers, physicists, meteorologists, sociologists, mathematicians, historians. It is up to the leaders in the field of emerging viruses, and perhaps up to writers like me who observe them at work, to make sure that this intermingling leads to an intellectual melting pot rather than to a jargon-ridden Tower of Babel.

The most basic question any virologist asks is this: How does a virus work? The question itself can be interpreted in many ways. How does a virus infect a host animal? How does it evade the surveillance of the host's immune system? How does it get inside a cell and force the cell to make more viruses? Each of these questions approaches the matter of viral functioning from a slightly different perspective. And each can be answered from a slightly different viewpoint: that of either the genes of the virus, the environment of the host cell, the nature of the host animal as a whole, or the relationship between the host animal and the world in which it moves. In other words, questions about viral functioning can be answered from the subdisciplines of, respectively, molecular biology, cell biology, clinical infectious disease, and population biology.

In the 1990s, the viewpoint of molecular biology seems to hold all other perspectives in its thrall. "Biology has been transformed by the ability to make genes and then the gene products to order," writes Walter Gilbert, an emeritus biologist at Harvard. "Developmental biology now looks first for a gene to specify a form in the embryo. Cellular biology looks to the gene to specify a structural element. And medicine looks to genes to yield the body's proteins or to trace causes for illnesses. Evolutionary questions—from the origin of life to the speciation of birds—are all traced by patterns on DNA molecules. Ecology characterizes natural populations by amplifying their DNA. The social habits of lions, the wanderings of turtles, and the migrations of human populations leave patterns on their DNA." The ascendancy of molecular biology, Gilbert concludes, has amounted to a "paradigm shift" in biology—a phrase originally used by science historian Thomas S. Kuhn to describe a change in the fundamental premises that guide research in a field. Even though much of the day-to-day lab work of gene cloning is both tedious and restrictive, biologists seem chained to the notion that any question they investigate must be explored at the level of the gene. Biologists often complain that "sequencing is boring," writes Gilbert. "And yet everyone is sequencing." They are doing so in self-defense, ever aware that other biologists will not take their work seriously unless it is based on identifying, cloning, sequencing, and expressing relevant stretches of DNA.

This paradigm shift essentially amounts to a sharp turn toward reductionism—the process of stripping things down to their tiniest intact essence and studying only that small component. In chemistry, reductionism ultimately leads to the atom; in biology, to the gene. Molecular biologists believe they can answer just about any interesting question in biology by hunting down the gene: splitting apart the double helix of DNA, snipping off sections to figure out which nucleotide goes where, copying stretches of genetic material hundreds and thousands and millions of times over. They believe that virtually any biological phenomenon, from cell division to pathogenesis to altruistic behavior, can be explained by an organism's DNA. It's an exciting prospect, but it probably gives unwarranted weight to the action of the genes. This reductionistic view of biology ignores so many other factors—at the level of the cell, the organism, and the environment—that it is of limited relevance in trying to understand how nature really works. Because of its shortsightedness, and its focus on the tiny, I believe genetic investigation cannot quite measure up to the elevated status it has attained in the world of biology. It seems a case of trying to turn a tool into a philosophy.

Molecular biologists are the intellectual heirs of the tinkerers of a few generations back, those folks who spent all their spare time in the garage or the barn, taking things apart and reconfiguring them into marvelous new machines. Like those early inventors, molecular biologists like to tear things down to their most basic component—the gene—hoping to arrive at a more refined, more specific analysis of how things work. Then they try to piece everything back together in interesting and creative new ways. To elevate these tinkerers to the rank of master journeyman is probably too great a leap.

"Why are we doing all this sequencing?" asks D. Carleton Gajdusek of the National Institutes of Health. "The young people in my lab tell me they're collecting interesting data. But is it sensible data?" Gajdusek says that the ability to sequence a particular protein is not unlike making a plaster cast of a tree; it allows man to mimic nature precisely. But lost in the excitement of doing it because it *can* be done is any notion of whether it *should* be done. It's one thing, says Gajdusek, to make a plaster cast of a dwarf tree growing in a greenhouse in order to sell the copies as decorations during the Christmas season; in that case, there's a utility to the exercise. But it's quite another thing to set about making plaster casts of all the trees in the forest, with no ultimate goal in mind. "We are making very good copies of proteins, but to what end?" he asks. "I'm beginning to think it's all an infinity of wasted time."

Like Gajdusek, many of biology's most venerable researchers dismiss some experiments in molecular biology as "cookbook science," requiring little more sophistication, technically or intellectually, than is required to follow a recipe. Because of commercial reagents now available to speed along the splitting of the DNA chain, and because of computer programs that do the letter-by-letter analysis for you and compare them to genes on a master database, sequencing can be accomplished by any biology undergraduate with steady enough hands to measure quantities accurately. Even I have successfully sequenced genes (under the watchful guidance of some very patient lab instructors) during a summer course for science writers, despite the fact that I had very little idea of precisely what I was doing or why. If sequencing is something even a journalist can do, how can it be the holy grail so many biologists treat it as?

This is not to dismiss the value of molecular genetics as a fabulous new approach to research. It's truly astounding when a scientist can associate a change in a cell or an organism with parallel changes in a few blips on a chromosome. The caveat is this: there are limits to this kind of work. Scientists must remember—as most of them do—that figuring out the genetics of an association is only the beginning of their investigation, not the end.

Is this sanctification of the gene good or bad for biology? It has its advantages, no doubt: an emphasis on genetics can offer biologists a clarifying concept that imposes order on seemingly unrelated events. But I think the tendency toward reductionism is, on balance, bad for the discipline. It means most biologists are looking at smaller and smaller pieces of smaller and smaller questions. Their sights can become so myopic that they ultimately have no understanding of, or even interest in, how their object of study fits into the larger picture. "I have a friend who's studying retroviruses," says Morse, who despite his relatively young age still has an old-fashioned fascination with pathogenesis. "She doesn't even think of herself as a virologist; she thinks of herself as a molecular biologist who happens to work with viruses."

The myopia of reductionism can be blinding. By peering too closely at viruses, zooming in on the smallest possible unit of life, the biologist gets a distorted view of their differences and their similarities. Differences at this scale loom large, and the scientist is unable to take a few steps back and recognize commonalities among one particular virus and another. Nowhere is this more apparent than in the case of poliovirus. Long before the structure of the poliovirus was understood, medical researchers knew there were three types of polio, and that infection with one type did not protect an individual from subsequent infection with either of the other two. Without any understanding of the RNA that was in the virus, scientists were able to make an effective polio vaccine that immunized against all three types. But once molecular biologists examined the poliovirus more closely, the differences among the three types emerged clearly. There turned out to be hundreds and hundreds of slightly different strains, even among viruses that functionally belonged to the same type (in terms of the class of antibodies produced in response to the most important antigens).

"It's like saying that these are all elm trees, but if you look a little closer you can see that every elm is a little bit different," says Gajdusek. As far as the disease itself, the differences among these hundreds of strains are of no consequence; all that matters is that type 1 poliovirus leads to one antibody reaction, type 2 to another, and type 3 to a third. From the standpoint of public health, only those three types are relevant. Immunize against all three, and a child is protected for life. But molecular biologists, peering at their poliovirus sequences and their crystallographic records of its structure, cannot know this. They see the trees, not the forest. If polio were being researched today, Gajdusek says, scientists would begin with the molecular structure of the virus and see the differences before they see the commonalities. "They would say, 'Forget about this; they're all different. There's no way we can make a vaccine against them all.'" That may be one of the problems stymieing today's development of a vaccine against HIV infection: with so much emphasis on the differences among strains, even sometimes within the same individual, scientists are having a hard time categorizing the virus into types that are of biological significance— types that relate to the development of classes of antibodies that confer immunity to many strains of HIV.

What's especially nice about the study of emerging viruses is that it does not allow for such reductionism to take hold too strongly. People working at the cutting edge of their fields cannot afford to be limited in their scope. At the frontier, the only thing visible is the broad horizon. "Most scientific nonreductionists," wrote Keith Stewart Thomson, a biologist at Yale, in 1984, "get there by finding the limitations of applying simple reductionist strategies to complex systems and through the discovery of 'emergent' phenomena." It's a propitious turn of phrase that he mentions "emergent phenomena" a full five years before the term emerging viruses even came into formal use. But it's no accident that he saw the link between an interdisciplinary orientation and activity at the cutting edge of biology—the same connection that I see now in relation to emerging viruses. No matter what new problem is being investigated, biologists cannot do justice to it until they get past the artificial boundaries of disciplinary allegiances and self-imposed reductionism.

"My article of faith is that the questions of virology are interdisciplinary questions," says Bernard Fields, chairman of microbiology at the Harvard Medical School. "The only way to understand the genetics and molecular biology of pathogenesis is to study it at the animal level. You can't study viruses just in cell culture. There are host-specific factors that you'd never know about otherwise, that might give you new ways to think about viruses." Fields cites the example of the reovirus, which he has spent his entire professional life studying. The traditional view of any virus, he says, is that it is entirely inert until it gets into a cell. If reovirus had been studied only in cell culture, that view would have persisted. But by looking at how the virus works in an animal, Fields uncovered something unusual about the virus's ability to convert from an inactive to an active state, even before it infects its first cell. In the lumen of the intestine—the space inside the long coiled intestinal tube, which is still really part of the body's external environment rather than part of the body itself—an enzyme initiates a series of rearrangements that changes the virus from a sporelike form to an activated form. Only when it is activated is the reovirus capable of crossing from the lumen into the intestinal cells. "So you swallow the virus," says Fields, "it goes into the lumen of the intestine, and in the lumen the virus is converted to a form that crosses these cells and then starts multiplying in the primary site. How are you going to learn this in cell culture?"

By interdisciplinary, Fields means cutting across the subdivisions within the general field of biology. But the study of viruses should be even more expansive, encompassing perspectives from totally different disciplines. We've already seen how different factors come into play in the evolution of most new viruses. Think of arboviruses: it's not enough to know about the molecular genetics of the virus itself. To really stop the movement of an arboviral disease into a new population, you need the involvement of experts in entomology, meteorology, aquatics, political science, anthropology, veterinary sciences, even economics.

Gajdusek bemoans the fact that his own lab associates have lost the grasp of geology and physics that is the legacy of any curious third-grader fascinated with phenomena of the natural world. They have forgotten, he says, their earliest science "experiments," allowing salt crystals to form on a piece of rock salt suspended in solution, observing the way stalactites and stalagmites grow in caverns and actually come to resemble each other, building bridges and arches with Popsicle sticks or wooden blocks. "I'm having trouble getting molecular biologists to understand what they knew in elementary science—that even salt and water can have an infinite number of patterns," says Gajdusek, whose own extracurricular study of linguistics has helped him formulate some of his most original ideas about how viruses work. He wants his younger associates to remember some basic physics and chemistry—patterning, for example, and crystallization—because he thinks these processes can be useful models for explaining how agents involved in transmissible brain diseases such as kuru, Creutzfeldt-Jakob disease, and bovine spongiform encephalopathy conduct their own replication. Too emphatic a focus on genes would limit insight into these "unconventional viruses," he says, because not one of them contains any RNA or DNA.

The argument among laboratory scientists over whether too much emphasis is placed on genetics is nothing compared to a more vitriolic argument going on concurrently within biology: the turf battles between lab scientists and fieldworkers. Most contemporary biologists spend their days inside a laboratory, mixing things and using sophisticated equipment. But field biologists, who are notorious for their machismo, dismiss these endeavors as too tame; they think of their own exploits as more akin to those of Indiana Jones. They tend to denigrate the scientists who work in high-containment laboratories and put on blue rubber space suits before handling lethal viruses. Isn't risk taking a part of the scientific enterprise? Weren't old khakis and a pair of rubber gloves protection enough for them, working through dust, glare, and sweat in the bush of Africa or the forests of South America?

"Field virologists still hold the informal 'Walter Reed club' in high regard—the roster of fellow scientists who have died in the line of duty," says Tom Monath, whose previous work, with the U.S. Army Medical Research Institute of Infectious Diseases, had taken him to the field in Central and South America. This "club" is named in honor of Major Walter Reed, the tropical disease expert after whom the army's medical research facility in Washington, D.C., is also named. In 1900, the Reed Commission was sent to Cuba to investigate an outbreak of yellow fever, whose transmission pattern was at the time still unknown. By using human subjects, who volunteered to be bitten by mosquitoes that had just fed on yellow fever victims, the commission proved that the disease was carried by mosquitoes. But this demonstration came at a high cost. Reed and several of his co-workers volunteered to be bitten. A few—though not Reed—developed yellow fever and died.

Fieldwork today is not quite so risky as it was then. But even though it no longer involves being deliberately bitten by virus-carrying vectors, it remains a true test of mettle. It involves living under primitive conditions, working in laboratories where the electrical generator goes out regularly, scrambling for enough liquid nitrogen to keep samples cold, training natives in techniques of animal trapping or basic microscopy, bartering for supplies, negotiating with government authorities, and waiting, waiting, waiting. This is certainly not the lifestyle envisioned by today's young physician or new biology graduate, two academic paths that have traditionally helped fill the ranks of field virologists.

In the 1960s and 1970s, when the United States had a military draft, thousands of doctors signed up for two-year stints with the Public Health Service as an alternative to military service—and this proved to be a wonderful way to convince the nation's best new doctors of the charm of fieldwork in international health. Every year, thirty-five young men enlisted with the Centers for Disease Control's Epidemiology Intelligence Service. "We had to stop taking applications when it got to four or five hundred," recalls D. A. Henderson, who worked at the CDC at the time. This meant that those who were finally accepted were the best of the brightest. After two years in the field, often with an overseas assignment, many of these young doctors got "converted" to international public health, Henderson says. "They had intended to be internists or cardiologists; they hadn't thought about public health. But they saw the potential of public health service in the United States through the perspective of health care overseas." Unfortunately, there is no such required service these days. The Epidemiology Intelligence Service still exists, and it still sends out young physicians to investigate perplexing disease outbreaks wherever they occur. But the men and women engaged in this work come to it already converted. Most young doctors don't feel that they have time to rough it before settling down to a private practice or a research career. Going through medical school typically involves incurring a debt as high as $100,000, and more and more new M.D.s are choosing their specialties based on the income they can expect as payback for that enormous financial burden. So the very brightest young physicians go into suburban practices in orthopedics or radiology, which promise limited hours and respectable incomes.

Young biologists, too, are finding that the professional rewards they have worked so hard for—research grants, promotion and tenure, status among their peers—are not generally best pursued by spending time working in Africa or Asia. They are following the money and the prestige and choosing to stay in the laboratory to sequence genes. Federal and private support for field research, which is quite expensive, has all but disappeared. What research money is available these days is scarce, and it is almost all routed toward traditional laboratory science, especially science with a molecular component.

These trends are worrisome to those of us concerned about controlling emerging viruses. The only way to track the course of new viruses is with a greater emphasis on fieldwork, not on cloning. For any of the "listening post" ideas now being floated about how to anticipate the next virus, field virologists play a central role. But where will those field virologists come from? I expect they will be recruited not from biology departments, where the gene is king, but from departments where people have already been shown to have a fascination with fieldwork—even if it's not necessarily fieldwork related to public health. Tomorrow's field virologists are probably majoring today in ecology, anthropology, marine biology, geology, entomology, agriculture, geography, maybe even political science or sociology. These people have already developed a taste for work in the field—a highly idiosyncratic taste, involving as it does discomfort, inconvenience, and frustration—and need only a nudge in the right direction to see how their inherent interest in the natural world can be applied to investigations into human health and disease. When they are ready to pursue graduate study, that is the time to offer them inducements in the form of fellowships and grants to conduct work in international health, and to offer them additional training in some basic biology so they can apply what they know to biological questions. Indeed, a recent report from the National Academy of Sciences on "Opportunities in Biology," which emphasized the importance of an interdisciplinary approach to fieldwork, suggested just that. Encourage young people "to do predoctoral training in one field and postdoctoral in another," the report concluded. A great idea—especially if "encouragement" translates into money. In a scientific culture that emphasizes differences rather than continuities, bridging the gaps that separate the disciplines is a risky venture. No ambitious student should be expected to take that risk without adequate financial compensation.

Big obstacles stand in the way of the "new biology." Disciplinary barriers are entrenched in the American academic structure: most universities are run as federations, with individual departments making most of the decisions about graduation requirements, faculty appointments, promotion and tenure, and long-range planning. To some extent, this has been useful. It helps colleagues talk to one another; it allows them to take for granted certain ideas, methods, and vocabulary. Each discipline uses its unique perspective to help its proponents get a grasp on one small corner of the world. But when disciplinary specialization becomes too rigid, as has happened throughout academia, the reliance on certain pre-existing assumptions can be crippling instead of liberating. People trained with a narrow view of one perspective tend to get stuck intellectually in their own old habits. Even their first questions—What is worth looking at? What do we wonder about?—are grounded in everything that went before, and the kinds of explanations on which they rely to answer these questions are stale as well.

At one point, Stephen Morse toyed with the idea of forging new bridges across disciplines by creating a new discipline of his own. Several of his colleagues were taken with his "viral traffic" metaphor to describe the way new viruses emerge. One of them—Daniel Fox, a sociologist and president of the Milbank Memorial Fund, a health care philanthropy in Manhattan—actually encouraged Morse to consider expanding the metaphor into a whole new interdisciplinary field of study, to be called "traffic science." The way Fox and Morse envisioned it, traffic science could cut across departmental lines and bring in expertise from a wider range to address a common problem. But in the end, both of them realized that creating a new subdiscipline was counterproductive. "I don't want a new department, with its own journals and meetings and jargon," Morse says. "I'd like to break down department barriers and get people to cross disciplines, not by inventing new departments, but by inventing ways to get everyone together in the same cafeteria."

Interdisciplinary thought in biology can be encouraged in a few ways. One is to take a cue from medicine. When it became clear, in the 1970s, that the competition to get into medical school was forcing students to concentrate exclusively on science courses, medical school deans began to worry. They found their applicant pool overflowing with straight-A students, but they worried about the breadth of knowledge these science majors would bring to the portion of medicine that is an art rather than a science. To get around the prospect of turning out a generation of narrow-minded physicians, some medical schools began actively recruiting applicants from nontraditional majors: philosophy, literature, history. The same can be done in biology. If Carleton Gajdusek is worried that the postdocs in his laboratory don't really understand why they are performing the elaborate party tricks of molecular manipulations, perhaps he should be looking for postdocs who have a strong background in the humanities rather than the sciences.

The next step in encouraging interdisciplinary cooperation is a bit more informal. It goes back to Morse's idea of getting people from different departments "together in the same cafeteria." His inspiration for this way of viewing the world, as well as my own inspiration, comes from the time we each spent in the near idyll of the Marine Biological Laboratory (MBL) in Woods Hole, Massachusetts. Morse spent several summers there in the 1980s as a scientist conducting his own research; I spent four weeks there in 1990 as a science writer watching biologists at work. (Sometimes I even helped out; it was at Woods Hole, for instance, that I managed to sequence DNA.) At the

MBL, biologists from across the country rent small laboratories or library carrels for the summer and congregate in the halls of the buildings, on the bike paths or the beaches, or in the two pubs on the three-block-long "downtown." In this casual way, a physiologist from Chicago might find out what an evolutionist from Boston—or from Bonn—is working on, and each might have ideas the other never would have thought of. The Woods Hole experience is so profound for biologists that many of them have written rhapsodic essays about the fertility of the interdisciplinary approach. "You can hear the sound from the beach at a distance, before you see the people," writes Lewis Thomas, former president of the Memorial Sloan-Kettering Cancer Center and a member of the board of seven hundred scientists that administers the Marine Biological Laboratory. "It is that most extraordinary noise, half-shout, half-song, made by confluent, simultaneously raised human voices, explaining things to each other." If the MBL ambiance can be created in other environments, with scientists from many subdisciplines interacting over a prolonged period in an informal setting, interdisciplinary cooperation could well take an important step forward.


Viruses gradually (or sometimes suddenly) change, and occasionally they turn up in places where they've never been before. But the basic lesson of the study of emerging viruses is that weird new strains of pathogens are not waiting in the wings ready to pounce; the odds are against having all factors converge at one moment in time to allow a new virus to emerge. Even knowing this, though, it would be folly to grow too complacent about the security of mankind's place in the biosphere. AIDS made us all too aware of that. As Mathilde Krim, a professor of public health and founder of the AIDS Medical Foundation, said recently, "AIDS is giving us a lesson in humility. It teaches us that the relentless evolutionary forces at play in nature continue to create new forms of life—such as new strains of viruses—and that for all mankind's arrogance and destructive powers, we are not yet the masters of the universe, nor even necessarily nature's most favored creatures."

Maybe we're not the most favored, but we're still among nature's most creative creatures. The challenge facing us in the next century will be to apply what we have learned about the basic blueprint of viral emergence to anticipate the next new plague, and do our best to defang it before it devours us. This will require some humility, there is no doubt. But it will also require a healthy respect for our own intellectual prowess and a willingness to develop a united front, a coalition of all our scientific resources, to do silent battle with mankind's tiniest and perhaps most complicated foe.

# Acknowledgments

Dozens of scientists contributed their time, their expertise, and their extraordinary patience during the course of my research for this book. I am grateful to them all, but especially to Steve Morse, who was forever gracious—and willing to share every bit of relevant information in his possession—over many months and many meetings. His insight into all aspects of virology, as well as into history, politics, mathematics, and even classical music, was a continual source of delight to me, as was his willingness to review sections of the manuscript for scientific accuracy. Others who reviewed portions of the manuscript also deserve thanks: Robert Shope, Bernard Fields, Thomas Monath, Joseph Melnick, W. French Anderson, Max Essex, Edwin D. Kilbourne, Anthony Komaroff, Richard M. Krause, Ian Lipkin, Colin Parrish, Dick Montali, Susan Fisher-Hoch, Brian Mahy, and Andrew Spielman. If errors are to be found in the book, the responsibility for them is mine; but if errors have been avoided, that owes no small measure to the assistance of these generous folks.

This book began as the inspiration of Michele Slung, a fellow writer who has such an innate sense of the good story that she saw one in viruses even as she was hospitalized with an exotic viral ailment. Through Michele, I eventually found my way to Jonathan Segal of Knopf and Marty Asher of Vintage, two intelligent and sensitive men who took a chance on this exciting topic. My heartfelt thanks to them, as well as to my determined agent, Barbara Lowenstein.

## Acknowledgments

And finally, a thank-you for my family, who helped me live a normal life as I worked my way through a project that at times seemed overwhelming. To my husband, Jeff, as ever my first and best reader, my mate in every sense of the word; and to our daughters, Jessie and Sam, I can say only this: you are all wonderful.

# Notes

## Introduction

page ix. By mid-October 1993, when this book went to press, epidemiologists from the Centers for Disease Control and Prevention (CDC) had identified cases in several states beyond the original outbreak. Confirmed or suspected cases of infection with what they were tentatively calling the Four Corners Virus had spread to California, Louisiana, Montana, Nevada, North Dakota, and Oregon. At press time, the running toll of the disease was 55 suspected cases, 39 confirmed, and 23 deaths. For more information, see Denise Grady, "Death at the Corners," *Discover*, December 1993, page 82; Natalie Angier, "In Navajo Land of Mysteries, One Carries a Deadly Illness," *New York Times*, June 5, 1993, A1, and "Tracks of Mystery Disease Lead to New Form of Virus," *New York Times*, June 11, 1993, A14; David Cannella, "Medicine Men Found Answers in the Trees," *Washington Post*, June 6, 1993, A9; Sue Anne Pressley, "Navajos Protest Response to Mystery Flu Outbreak," *Washington Post*, June 19, 1993, A3; and Bernice Wuethrich, "Army Scientists Isolate Deadly Virus," *Science News*, August 21, 1993, 116. Also see the series of articles, all entitled "Update: Outbreak of Hantavirus Infection—Southwestern United States, 1993," published by CDC in the *Journal of the American Medical Association*, vol. 270: July 21, 1993, 306; July 28, 1993, 429; August 25, 1993, 934; September 8, 1993, 1176; and October 27, 1993, 1920.

xii. The declaration of victory in the war against infectious disease made by Surgeon General Stewart was paraphrased in Donald A. Henderson, "Surveillance Systems and Intergovernmental Cooperation," in Stephen S. Morse, ed., *Emerging Viruses* (New York: Oxford University Press, 1992). Sir Macfarlane Burnet made his own similar claims in the third edition of his book, *The Natural History of Infectious Disease* (London: Cambridge University Press, 1962), 3.

*Notes*

page xii. The "colonies of viruses" image is from Richard Dawkins, *The Selfish Gene*, 2nd ed. (Oxford, England: Oxford University Press, 1989), 182.

xiii. Richard M. Krause's comments on microbial emergence were made in "After AIDS: The Risk of Other Plagues," *Cosmos*, Fall 1991, 15–21.

xiv. Details about the viral impact of the Midwest flooding of 1993 are in Paul Cotton, "Health Threat from Mosquitoes Rises as Flood of the Century Finally Recedes," *JAMA*, 270, no. 6 (August 11, 1993): 685–686. The early outbreak of Beijing flu is described in an Associated Press article, "Flu Outbreaks Worry Health Officials," *Washington Post*, September 13, 1993, A9.

xv. The two scholarly works whose publication roughly coincided with hard-cover publication of this book are Stephen S. Morse, ed., *Emerging Viruses* (New York: Oxford University Press, 1992), and Joshua Lederberg, Robert E. Shope, and Stanley C. Oaks, Jr., eds., *Emerging Infections: Microbial Threats to Health in the United States* (Washington, D.C.: National Academy Press, 1992).

The prospective studies of emerging viruses in Papua New Guinea and Brazil were described in Ann Gibbons, "Where Are 'New' Diseases Born?" *Science* 261 (August 6, 1993): 680–681.

xvi. Joshua Lederberg has issued his dire predictions in many articles, most prominently in "Medical Science, Infectious Disease, and the Unity of Humankind," *JAMA* 260 (1988): 684–5.

Stephen Morse's quote is from his article "Regulating Viral Traffic," *Issues in Science and Technology*, Fall 1990, 81.

xvii. Krause's quote comes from his collection of essays, *The Restless Tide: The Persistent Challenge of the Microbial World* (Washington, D.C.: The National Foundation for Infectious Diseases, 1981), 26.

## 1. Why New Viruses Emerge

3. Details of the case about the man who brought Lassa fever home with him from Nigeria came from Lawrence K. Altman, "When an Exotic Virus Strikes: A Deadly Case of Lassa Fever," *New York Times*, February 28, 1989, C3.; Gary P. Holmes, Joseph B. McCormick, Susan C. Trock, et al., "Lassa Fever in the United States: Investigation of a Case and New Guidelines for Management," *New England Journal of Medicine*, October 18, 1990, 1120–3; and a personal interview with Joseph McCormick in May 1991.

7. The introduction of myxomatosis into Australia is considered a classic example of what happens in the first wave of an emerging virus disease. It was best described in Frank Fenner, "Biological Control, as Exemplified by Smallpox Eradication and Myxomatosis," *Proceedings of the Royal Society of London* 218 B (June 22, 1983): 259–85; and in Macfarlane Burnet, *Natural History of Infectious Disease*, 3rd ed. (Cambridge, England: Cambridge University Press, 1962), 195–7.

Estimates about the length of a rabbit generation are from William McNeill, *Plagues and Peoples* (New York: Anchor Press/Doubleday, 1976), 58; estimates about

236

the number of generations needed for virus and host to coevolve come from the same book, page 170, and from Max Essex and Phyllis J. Kanki, "The Origins of the AIDS Virus," *Scientific American*, October 1988, 64–71.

The factors necessary for the emergence of new viruses are itemized in Stephen S. Morse and Ann Schluederberg, "Emerging Viruses: The Evolution of Viruses and Viral Diseases," *Journal of Infectious Diseases* 162 (1990): 1–7.

9. The quotation from Lewis Thomas is from his book, *The Lives of a Cell: Notes of a Biology Watcher* (New York: Bantam Books, 1974), 89.

10. William McNeill's quote is from his book, *Plagues and Peoples* cited above, page 208.

11. The introduction of yellow fever into the New World was described by McNeill in *Plagues and Peoples* on page 213.

12. Daniel Fox, president of the Milbank Memorial Fund, New York City, made the remark about Morse in an interview in March 1991. He offered this description with a laugh, adding, "I love the guy."

16. Among the many articles that appeared in the professional press about the emerging viruses conference, some of the most lucid and thorough were Julie Ann Miller, "Diseases for Our Future: Global Ecology and Emerging Viruses," *Bio-Science* 39, no. 8 (September 1989): 509–17; Stephen S. Morse, "Regulating Viral Traffic," *Issues in Science and Technology* (Fall 1990): 81–4; Rick Weiss, "The Viral Advantage," *Science News* 136 (September 23, 1989): 200–3; and Mitchel L. Zoler, "Emerging Viruses," *Medical World News*, June 26, 1989, 36–42. The most thoughtful articles in the popular press that resulted from the conference included Lawrence K. Altman, "Fearful of Outbreaks, Doctors Pay New Heed to Emerging Viruses," *New York Times*, May 9, 1989, C3; John Langone, "Emerging Viruses," *Discover*, December 1990, 63–8; Kathleen McAuliffe, "The Killing Fields: Latter-Day Plagues," *Omni*, 1990, 51–4; and Stephen S. Morse, "Stirring Up Trouble: Environmental Disruption Can Divert Animal Viruses Into People," *Sciences*, September 1990, 16–21.

17. Morse and Ann Schluederberg co-authored a synopsis of the emerging virus conference that appeared as "Emerging Viruses: The Evolution of Viruses and Viral Diseases," *Journal of Infectious Diseases* 162 (1990): 1–7. The conference proceedings were collected in a book called *Emerging Viruses* (New York: Oxford University Press, 1992).

19. Edwin D. Kilbourne of the Mt. Sinai School of Medicine says that genetically deviant viruses fail to emerge into real disease threats because of the constraints imposed on their successful survival. These constraints are:

• Extreme genetic alterations are lethal to the virus.
• In order for the virus to survive, it must attain a high enough level of virulence.
• When viruses are propagated in the laboratory, in cell cultures made from animals that the virus does not normally even infect, the tendency is for the mutations to drive the virus in a less lethal (attenuated) direction.
• Evolutionary adaptation to ecological niches requires highly specific genetic changes.

- Population immunity is often so pervasive that it can be avoided only by a major change in the virus's genetic makeup—a change that is less likely to occur, either in the laboratory or in nature.
- Human infection with animal viruses, which does occasionally occur, is rarely contagious to another human.

Kilbourne expanded upon these ideas in his article "Epidemiology of Viruses Genetically Altered by Man—Predictive Principles," *Banbury Report 22: Genetically Altered Viruses and the Environment* (Cold Spring, N.Y.: Cold Spring Harbor Laboratory, 1985), 103–17.

Stephen Morse first made reference to his "viral traffic" metaphor in a chapter he contributed to a book edited by Daniel Fox and Elizabeth Fee, *AIDS: Contemporary History* (Berkeley: University of California Press, 1992). The chapter was called "AIDS and Beyond: Defining the Rules for Viral Traffic." At Fox's urging, Morse expanded the metaphor and gave it more prominence in another article "Regulating Viral Traffic," which appeared in *Issues in Science and Technology*, cited above.

page 21. Descriptions of the epidemic of Korean hemorrhagic fever among United Nations troops, and the epidemiology of the disease today, appear in James W. LeDuc, "Hantaviruses Model of Emerging Agent," *U S Medicine*, August 1990, 41–2.

23. A brief chronology of the emergence of Lyme disease appears in Thomas J. Daniels and Richard C. Falco, "The Lyme Disease Invasion," *Natural History*, July 1989, 4–10.

24. McNeill's statement is from *Plagues and Peoples*, page 17.

25. Information about global warming estimates, as well as the quote from George Craig, are from Marshall Fisher and David E. Fisher, "The Attack of the Killer Mosquitoes," *Los Angeles Times Magazine*, September 15, 1991, 30–5. Also quoted in the article was a World Health Organization report on the potential health effects of climatic change, which concluded, "It should be clear that climatic change will affect the distribution and prevalence of vector-borne diseases dramatically, and that these changes cannot be ignored."

26. Richard Levins's explanation of degree-days and the synchrony between predator and prey is from a personal interview conducted in March 1991.

27. Studies about the relationship between ultraviolet radiation and immune system functioning are summarized in Michael J. Lillyquist, *Sunlight and Health* (New York: Dodd, Mead, 1985), 115.

Thomas Lovejoy's quote is from page 515 of Julie Ann Miller's article "Diseases for Our Future," cited above.

The description of Argentinian hemorrhagic fever is from Karl M. Johnson, "Arenaviruses," in Alfred S. Evans, ed., *Viral Infections in Humans: Epidemiology and Control*, 3rd ed. (New York: Plenum Medical Books Company, 1989), 141.

28. Descriptions of how influenza viruses recombine in the "mixing vessels" of pigs, as well as Robert Webster's quote, come from Peter Radetsky, *The Invisible*

*Invaders: The Story of the Emerging Age of Viruses* (New York: Little, Brown, 1991), 246.

30. Edwin D. Kilbourne made his comments about viral mutation in immuno-suppressed patients during an interview in March 1991.

Clarence Gibbs made his comments about Creutzfeldt-Jakob disease in a lecture, "Conventional and Unconventional Virus-Induced Disorders in the Central Nervous System," presented to summer students at the National Institutes of Health, June 20, 1991.

31. Accounts of transmission of the Creutzfeldt-Jakob virus via corneal transplant and electrode contamination were recorded in P. Duffy, J. Wolf, G. Collins, et al., "Person-to-Person Transmission of Creutzfeldt-Jakob Disease," *New England Journal of Medicine* 299 (1974): 692–3.

32. The chronology of the Marburg outbreak is detailed in Karl M. Johnson, "African Hemorrhagic Fevers Caused by Marburg and Ebola," in Alfred S. Evans, *Viral Infections of Humans: Epidemiology and Control*, cited above.

A description of subsequent outbreaks of Marburg virus appears on page 16 of Stephen Morse's article, "Stirring Up Trouble," cited above.

33. The story of the trip of Ebola virus–infected monkeys from the Philippines to a Virginia laboratory is detailed in D'Vera Cohn, "Scientists Trace Ebola Virus's Deadly Path," *Washington Post*, December 11, 1989, D1.

William McNeill's quote appears in Julie Ann Miller's article "Diseases for Our Future," cited above, page 511.

34. Richard Krause made his comment in "After AIDS: The Risk of Other Plagues," *Cosmos*, Fall 1991, 15–21, which he offered in manuscript form during a personal interview in March 1991.

The story of the last two laboratory accidents involving smallpox appeared in Abram S. Benenson, "Smallpox," in Alfred S. Evans, ed., *Viral Infections of Humans: Epidemiology and Control*, cited above, page 651. Current efforts to map the smallpox genome are described in David Brown, "Computers to Hold Vestiges of Smallpox," *Washington Post*, May 11, 1992, A3.

Stephen Morse tells the story of the transfer of England's last stockpiles of variola (smallpox virus) to the high-containment lab at CDC. Variola cultures were packed in dry ice, wrapped in Styrofoam containers, and sealed in layer upon layer of protective wrap. Then the package was driven to Heathrow Airport under police escort: several police cars traveling alongside the car with the variola, sirens blaring, long sections of the motorway cordoned off to keep clear any potential terrorist activity. "It was the kind of security you'd expect for a motorcade for the Queen," Morse says. But when the variola arrived in Atlanta, a lone CDC employee was waiting for it. He picked up the package at the baggage claim, tossed it in the trunk of his car, and drove off.

The plans of the World Health Organization to destroy all remaining vials of variola, slated for New Year's Eve, 1993, were reevaluated at the last minute by scientists who were having second thoughts about deliberately wiping out an entire species, even a devastating one. As one former CDC official put it, "it is reckless and presumptuous to destroy the virus, because you cannot be sure that there won't be a need for it someday." For more on this eleventh-hour debate, see Lawrence K. Altman, "Scientists Debate Destroying the Last Strains of Smallpox," *New York Times*, August 30, 1993, A1.

page 35. Mirko D. Grmek commented on the English Sweate in his book *History of AIDS: Emergence and Origin of a Modern Pandemic*, trans. Russell C. Maulitz and Jacalyn Duffin (Princeton, N.J.: Princeton University Press, 1990), 103. Other details of the Sweate's history appear in Berton Roueche, *The Medical Detectives*, vol. 2 (New York: E. P. Dutton, 1984), 194. And the Sweate was one of the diseases mentioned in Paul B. Beeson, "Some Diseases That Have Disappeared," *American Journal of Medicine* 68 (1980): 806–10.

Krause's quote about the evolutionary tide is from a 1978 speech that appeared in his collection of essays, *The Restless Tide: The Persistent Challenge of the Microbial World*, cited above, on page 26. Morse's writings on viral evolution are from his chapter of that name that appears in Joshua Lederberg, ed., *The Encyclopedia of Microbiology*, in press.

36. Edwin D. Kilbourne drew his tongue-in-cheek profile of the MMMV in "Epidemiology of Viruses Genetically Altered by Man—Predictive Principles," *Banbury Report 22: Genetically Altered Viruses and the Environment* (Cold Spring, N.Y.: Cold Spring Harbor Laboratory, 1985), 103–17.

## 2. Case Study: Why AIDS Emerged

37. Details about early AIDS patients all come from the most comprehensive book about the emergence of the human immunodeficiency virus: Mirko D. Grmek, *History of AIDS: Emergence and Origin of a Modern Pandemic*, trans. Russell C. Maulitz and Jacalyn Duffin (Princeton, N.J.: Princeton University Press, 1990). The story of Margrethe Rask appears on pages 28–9; the story of Claude Chardon on pages 26–7; and the story of Gaetan Dugas on pages 18–19.

38. Many of the details of the exploits of Gaetan Dugas, and of other early AIDS patients, can be found in Randy Shilts, *And the Band Played On: People, Politics, and the AIDS Epidemic* (New York: St. Martin's Press, 1987; Penguin Books, 1988). This quote of Dugas's appears on page 165 of the paperback edition.

40. Stephen Morse's quote about AIDS is from his chapter "AIDS and Beyond: Defining the Rules for Viral Traffic," in Daniel M. Fox and Elizabeth Fee, eds., *AIDS: The Making of a Chronic Disease* (Berkeley: University of California Press, 1992).

Richard Krause made his comments during a personal interview in April 1991.

41. Mirko D. Grmek, on pages 104–5 of *History of AIDS*, cited above, draws many parallels between syphilis in the late fifteenth century and AIDS in the 1980s: syphilis, like AIDS, is transmitted both sexually and from mother to child; its emer-

gence led to the closing of public baths and a change in sexual morality in Europe; it began as a virulent disease with a high mortality rate; it was passed back and forth from Europe to North America with an important focal point in Hispaniola, now known as Haiti and the Dominican Republic; and a mild form of syphilis is native to baboons and gorillas in Africa, just as a mild form of AIDS seems to exist in African green monkeys and sooty mangabeys.

A description of the U.S. policy restricting the immigration of HIV-positive individuals appears in Malcolm Gladwell, "U.S. Visa Policy Denounced at Global AIDS Conference," *Washington Post*, June 20, 1991, A3.

42. Grmek presents his theory on pages 158–61 of his book, *History of AIDS*, cited above.

Figures estimating the rates of HIV-2 infection in western Africa are from Max Essex and Phyllis J. Kanki, "The Origins of the AIDS Virus," *Scientific American*, October 1988, 70. Essex points out, in a letter to the author dated November 1991, that the central African nations of Angola and Mozambique, connected to Europe by Portuguese shipping channels, do have discernible rates of HIV-2 infection. In the United States, only thirty-one people were infected with HIV-2 as of August 1991, according to Amy Goldstein, "Maryland Finds Rare Form of AIDS Virus in 4," *Washington Post*, August 3, 1991, B3. Most of the U.S. cases, which were centered in Florida and the northeastern United States, involved people who had come from western Africa or had had sexual relations with West Africans.

43. The relationship among HIV-1, HIV-2, and SIV is outlined in Ronald C. Desrosiers, "HIV-1 Origins: A Finger on the Missing Link," *Nature* 345 (May 24, 1990): 288–9. Descriptions of the relationship between HIV-2 and SIV also appeared in a letter to the editor from A. Karpas, "Origin and Spread of AIDS," *Nature* 348 (December 13, 1990): 578.

44. Max Essex offered his theory that scientists inadvertently inoculated research monkeys with HIV during a personal interview in May 1991.

46. Theories about the possible movement of an immunodeficiency virus from monkeys into human beings were presented—taken in the order in which they are listed here—in G. Lecatsas, "Origin of AIDS" (letter), *Nature*, 351 (May 16, 1991): 179; Karpas, cited above; Andrew Scott, *Pirates of the Cell: The Story of Viruses from Molecule to Microbe* (New York: Basil Blackwell, 1987), 240; and Scott, page 241. Theories about the relationship between polio vaccination in the 1960s in Zaire (then the Belgian Congo), and the introduction of HIV into Africa, are described in Tom Curtis, "Did a Polio Vaccine Experiment Unleash AIDS in Africa?" *Washington Post*, April 5, 1992, C3. Curtis also describes a complementary theory that AIDS spread among gays in the United States in the 1970s who used double doses of polio vaccine (which they did not know was contaminated with HIV) as an experimental treatment for genital herpes.

48. Details about the change in demographics in Zaire are from Mirko D. Grmek's book, cited above, page 176.

49. The case of the American soldier who developed AIDS after a smallpox vac-

cination was cited in Mirko D. Grmek's book, with an original reference made to the *Times* (of London), May 11, 1987.

page 49. This story of how AIDS might have originated is, in essence, the one relayed by Richard Krause during a personal interview in March 1991.

Moriz Kaposi's description of his first five patients originally appeared in the article *"Idiopathisches multiples Pigmentsarkom der Haut."* It has been translated into English and published in *CA—A Cancer Journal for Clinicians* 32 (1982), 343–7. Mirko D. Grmek also offered descriptions of the earliest cases of Kaposi's sarcoma, on page 112 of his book, cited above.

50. The case of the English sailor who died in 1959 is elaborated upon in Lawrence K. Altman, "Puzzle of Sailor's Death Solved After 31 Years: The Answer Is AIDS," *New York Times*, July 24, 1990, C3. It was originally reported in the medical literature in G. Williams, T. B. Stretton, and J. C. Leonard, "Cytomegalic Inclusion Disease and *Pneumocystis carinii* Infection in Adults," *The Lancet* 2 (1960): 951–5.

51. A brief explanation of how PCR works can be found in Barnaby J. Feder, "Dispute Arises over Rights for Copying DNA," *New York Times*, September 18, 1991, D7.

The tragedy of the Norwegian family that succumbed to AIDS is described in Mirko D. Grmek's *History of AIDS* on page 130. Grmek wrote that the two older daughters were still free of HIV infection in 1990.

52. Robert R.'s case is described in Mirko D. Grmek's book, on page 124.

53. Grmek's quote is from page 151 of his book.

For a description of how HIV was disseminated through the blood supply, see Harry W. Haverkos, "Epidemiology of AIDS in Hemophiliacs and Blood Transfusion Recipients," *Antibiotic Chemotherapy* 38 (1987): 59–65. Grmek's quote is from page 38 of *History of AIDS*.

54. Information about hemophilia and about the manufacture of Factor VIII appears in Suzanne Fogle, "AIDS Hemophiliacs in Tough Court Battle," *Journal of NIH Research* 3 (July 1991): 46. Information about the scope of HIV infection among hemophiliacs is from Gina Kolata, "Hit Hard by AIDS Virus, Hemophiliacs Speak Up," *New York Times*, December 25, 1991, 7.

Details of the case of William Norwood appeared in an article by the Associated Press, "Organ Donor with AIDS Virus is Identified," *New York Times*, May 19, 1991, 21.

55. Fauci's comment about non-HIV AIDS cases, made at a CDC conference attended by about three hundred scientists and sixty journalists, was reported in Jon Cohen, "'Mystery' Virus Meets the Skeptics," *Science* 257 (1992): 1032.

56. Daniel Fox, president of the Milbank Memorial Fund, made the comment about the transition of AIDS from an acute disease to a chronic disease during a personal interview in March 1991.

Robert Gallo's experiments with SCID-hu mice and HIV are described in

Jean Marx, "Concerns Raised about Mouse Models for AIDS," *Science* 247 (February 16, 1990): 809. A more technical description appears in that same issue of *Science*, on pages 848–51, in the article by Paolo Lusso and his colleagues, "Expanded HIV-1 Cellular Tropism by Phenotypic Mixing with Murine Endogenous Retroviruses."

57. Robert Gallo's quote is from the article by Jean Marx, cited above.

## 3. A Virus Primer

58. Examples of the colorful metaphors that have been applied to viruses over the years can be found, respectively, in the title of Andrew Scott's *Pirates of the Cell: The Story of Viruses from Molecule to Microbe* (New York: Basil Blackwell Inc., 1987); in Elaine Blume Wilson, *At the Edge of Life: An Introduction to Viruses* (Bethesda, Md.: Department of Health and Human Services, National Institute of Allergy and Infectious Diseases, 1980), 11; and in the statement made by Sir Peter Medawar and his wife, Jane, as quoted in Michael B. A. Oldstone, "Viral Alteration of Cell Function," *Scientific American*, August 1989, 42. The cleverest of these metaphors, the one about viruses resembling wayward teenage offspring, was offered by Peter Radetsky in his book *The Invisible Invaders: The Story of the Emerging Age of Viruses* (New York: Little, Brown, 1991), 402.

59. Andrew Scott provides evidence of the odd, nonbiological characteristics of viruses in his book, cited above, pages 34 and 35.

60. Information about the relative sizes of various viruses is from John M. Dwyer, *The Body at War: The Miracle of the Immune System* (New York: New American Library, 1988), 12.

The history of Beijerinck's contribution to the notion of viruses, and of the reactions to his theory, is outlined in Peter Radestsky's book, cited above, pages 65–7.

62. Evidence of the unnerving invincibility of smallpox viruses can be found in the letter to the editor by P. D. Meers, "Smallpox Still Entombed?" *The Lancet* 1 (May 11, 1985): 1103.

The viral tricks involved in endocytosis are alluded to in Chris Raymond, "Deception Is Everywhere in Life and Not Always Bad, Researchers Say," *Chronicle of Higher Education*, February 27, 1991, A5.

62. "Suicide proteins" made by the cell under the direction of the poliovirus are described briefly in Andrew Scott, cited above, page 49.

65. Much of the section describing the genetic code is based on Andrew Scott's description in "Building-Blocks," Chapter 2 of his book *Pirates of the Cell*.

The letters A, T, C, and G stand for the nucleotides adenine, thymine, cytosine, and guanine.

67. In RNA, the nucleotide that replaces thymine, known by the shorthand U, is uracil.

# Notes

page 68. The quote from Andrzej Konopka is from Natalie Angier, "Biologists Seek the Words in DNA's Unbroken Text," *New York Times*, July 9, 1991, C1.

69. Reference to Francis Crick's term "central dogma" is made in Mirko D. Grmek, *History of AIDS: Emergence and Origin of a Modern Pandemic*, trans. Russell C. Maulitz and Jacalyn Duffin (Princeton, N.J.: Princeton University Press, 1991), 55. More details about the central dogma, and the ways in which RNA viruses short-circuit or contravene it, appear in Peter Radetsky's book, cited above, page 299.

A highly detailed explanation of the routes by which different classes of viruses make mRNA can be found in David Baltimore, "Expression of Animal Virus Genomes," *Bacteriological Reviews*, September 1971, 235–41. His analysis is presented for the layman in Andrew Scott, cited above, pages 52–4.

70. The early years of "Teminism" are summarized in Peter Radetsky's book, cited above, pages 300–2. A description of the discovery of reverse transcriptase is in the same book, page 314.

71. For a vivid description of specific and nonspecific immunity, see Andrew Scott's book, cited above, pages 86–7.

An explanation of the growth of T cells, and their functioning, appears in John M. Dwyer's book, cited above, page 33.

72. A discussion of the adverse effects of the inflammatory immune response appears in Andrew Scott's book, pages 104–5.

73. This brief outline of the origins of autoimmune disease is still controversial among immunologists. For more discussion, see John M. Dwyer's book, cited above, pages 38–9.

The role of antibodies is clearly elucidated in John M. Dwyer's book, cited above, on pages 93–5.

74. Regarding the speed of viral replication, Peter Radetsky, on page 9 of his book, cited above, was referring to a bacteriophage, a virus that infects bacteria. The rate of replication is thought to be similar for many animal viruses.

75. Information about the host range of the rabies virus appears in the chapter by Robert E. Shope, "Rabies," in Alfred S. Evans, ed., *Viral Infections of Humans: Epidemiology and Control* (New York and London: Plenum Medical Books, 1989), 509.

Bernard Fields explained the action of the rabies virus during a personal interview in May 1991.

76. A description of experiments with LCM and mice appears in C. A. Mims, *The Pathogenesis of Infectious Disease* (New York: Academic Press, 1982), 191–2.

David Baltimore made his comment about the meaning of life to Peter Radetsky, who quoted him on page 393 of his book, cited above.

Richard Dawkins's quote about viruses as "rebel human DNA" is from his book, *The Selfish Gene* (Oxford, England: Oxford University Press, 1989), 182.

77. James Strauss's theory about transposons and retrotransposons is explained in Peter Radetsky, cited above, pages 400–1.

244

Fred Hoyle makes his statements about the extraterrestrial origin of viruses in *The Intelligent Universe: A New View of Creation and Evolution* (New York: Holt, Rinehart and Winston, 1984).

78. Lewis Thomas's quotation about the "dancing matrix of viruses" is from *The Lives of a Cell: Notes of a Biology Watcher* (New York: Bantam Books, 1974), 4.

80. Russell Doolittle's comments are from his article in the "News and Views" section of *Nature*, June 1, 1987.

Mal Martin described his work on the human placenta gene to the author at a personal interview during the writing of "Viruses Revisited," *New York Times Magazine*, November 13, 1988.

81. More on computer viruses can be found in David Stang, "PC Viruses: The Desktop Epidemic," *Washington Post*, January 14, 1990, B3.

### 4. Mad Cows, Dead Dolphins, and Human Risk

85. The chronology of the Ebola virus importation has been in a series of articles by D'Vera Cohn in *Washington Post*: "Deadly Ebola Virus Found in Virginia Laboratory Monkey," December 1, 1989, A1; "Scientists Trace Ebola Virus's Deadly Path," December 11, 1989, D1; "Four Handlers in Virginia Get Ebola Virus," April 6, 1990, B5. An update of the story was provided by Susan Fisher-Hoch of the CDC in a letter written to the author in October 1991.

86. Fatality rates in the early outbreaks of Ebola fever are found in Karl M. Johnson, "African Hemorrhagic Fevers Caused by Marburg and Ebola Viruses," in Alfred F. Evans, ed., *Viral Infections of Humans*, 3rd ed. (New York and London: Plenum Medical Book Company, 1989), 97.

In early 1990, the CDC conducted a survey of 550 people whose jobs put them in close contact with monkeys. By then, four of the closely watched Hazleton lab workers had developed Ebola antibodies in their bloodstreams. In this larger survey, 266 people worked in facilities like Hazleton, and 284 had other forms of exposure to monkeys or their tissue. The study looked for Ebola antibodies in a control group of 449 randomly selected adults as well. The results showed that exposure to this new virus was more widespread than expected. Ten percent of the workers in animal quarantine facilities, 5.6 percent of workers exposed in different venues, and nearly 3 percent of ordinary Americans had antibodies to one of the four known filovirus strains: Ebola/Zaire, Ebola/Sudan, Marburg, and the new strain known as filovirus/Reston. But no one with antibodies ever got sick. For more information, see D'Vera Cohn, "Fearing Virus, New York Restricts Imports of Monkeys Used in Research," *Washington Post*, March 22, 1990, A15.

88. Stephen Morse quotes Francis Crick in his article "Emerging Viruses," *American Society for Microbiology News* 55, no. 7 (1989): 358. He made his comments about humans as accidental hosts for viruses during a personal interview in March 1991.

89. Jean-Michel Bompar was quoted in Marlise Simons, "Virus Linked to Pol-

lution Is Killing Hundreds of Dolphins in Mediterranean," *New York Times*, October 28, 1990, 3.

page 89. Figures for the total numbers of dolphins dying of seal plague, and the possible threat to monk seals, come from Alan Cowell, "A Poisoned Season: Dead Dolphins, Abused Pups," *New York Times*, September 4, 1991, A4.

90. Particulars about the spread of measles in populations are from Francis L. Black, "Measles," a chapter in Alfred F. Evans's book, cited above, pages 451 and 456.

92. Brian Mahy detailed his involvement in the seal plague mystery in a personal interview in August 1991, as well as in his chapter, "Seal Plague Virus," in Stephen S. Morse, ed., *Emerging Viruses* (New York: Oxford University Press, 1992).

Theories about where seal plague virus originated are found in C. B. Goodhart, "Did Virus Transfer from Harp Seals to Common Seals?" *Nature* 336 (1988): 21.

93. Robert Bullis's quotes, and other information about the death of Caribbean sea urchins, are from William Booth, "Mysterious Malady Hits Sea Urchins," *Washington Post*, August 11, 1991, A4.

94. Irene McCandlish's comments were quoted in an article, "Canine Parvovirus—A New Disease," in *Veterinary Record*, September 29, 1979, 292.

The next cases of canine parvovirus, also based on analysis of frozen sera, could be traced to Belgium in 1976, the Netherlands in 1977, Australia in May 1978, and Japan in July 1978. See Colin R. Parrish, "Canine Parvovirus 2: A Probable Example of Interspecies Transfer," a chapter in Stephen S. Morse's book, cited above.

Details about the life cycle of the parvovirus can be found in Andrew Scott, *Pirates of the Cell* (New York and London: Basil Blackwell, 1989), 55.

97. Colin Parrish described his work in a personal interview in September 1991, as well as in his chapter in Stephen S. Morse's book, cited above.

98. Parrish's comments discounting the theory that the feline distemper virus vaccine was the source of the recently emerged canine parvovirus 2 are from a letter written to the author in September 1991.

100. Diane Ackerman's quote about the golden lion tamarin is from her article "Golden Monkeys," in *The New Yorker*, June 24, 1991, 37. Details about the tamarin's mating habits are from the same article, page 38.

101. Statistics on the success rates of reintroduction campaigns are found in Christopher Anderson, "Keeping New Zoo Diseases Out of the Wild," *Washington Post*, June 3, 1991, A3.

Details about the case of Flash were provided by Richard J. Montali in a personal interview in September 1991.

102. The scenario about how CHV could spread through tamarin carcasses ap-

peared in Diane Ackerman's article, cited above, page 39, and was mentioned by Richard J. Montali as well.

Benjamin Beck made his comment during a personal interview in September 1991.

103. The film being described is *Warning Sign*, directed by Hal Barwood, produced by Jim Bloom, written by Hal Barwood and Matthew Robinson, starring Sam Waterston, Kathleen Quinlan, and Yaphet Kotto, released by Twentieth Century-Fox in 1985.

Neal Nathanson offered his disclaimer during a personal interview in May 1991.

104. Robert Gallo's work was described in Jean Marx, "Concerns Raised About Mouse Models for AIDS," *Science* 247 (February 16, 1990): 809.

The quotes from the article are from Paolo Lusso, Fulvia di Marzo Veronese, et al., "Expanded HIV-1 Cellular Tropism by Phenotypic Mixing with Murine Endogenous Retroviruses," *Science* 247 (February 16, 1990): 848, 851.

105. Comments from Goff and Feinberg appeared in Gina Kolata, "AIDS Expert Warns of Hazards in Research," *New York Times*, February 16, 1990, A19.

The four cases involved in the herpes B outbreak in Pensacola are described in detail by epidemiologists at the Centers for Disease Control in "B-Virus Infection in Humans—Pensacola, Florida," *Morbidity and Mortality Weekly Report* 36, no. 19 (May 22, 1987): 289–96.

107. The history and mortality rates of herpes B are from A. E. Palmer, "B-virus, *Herpesvirus simiae*: Historical Perspective," *Journal of Medical Primatology* 16 (1987), 99–130.

110. The possibility that archaeologists should be vaccinated against smallpox was raised in P. Meers, "Smallpox Still Entombed?" *The Lancet* 1 (May 11, 1985): 1103.

The chronology of the origins and development of "mad cow disease" in England was best described in the following articles: Richard H. Kimberlin, "Bovine Spongiform Encephalopathy: Taking Stock of the Issues," *Nature* 345 (June 28, 1990): 763–4; Jeremy Cherfas, "Mad Cow Disease: Uncertainty Rules," *Science* 249 (September 28, 1990): 1492; Constance Holden, "Antelope Death Adds to BSE Worries," *Science* 250 (November 30, 1990): 1203 and "Answer to Mad Cow Riddle?" *Science* 251 (March 15, 1991): 1310; and Peter Aldhous, "Antelopes Die of 'Mad Cow' Disease," *Nature* 344 (March 15, 1990): 183.

## 5. Do Viruses Cause Chronic Disease?

112. The description of the Marek's disease vaccine is from Elaine Blume Wilson, *At the Edge of Life: An Introduction to Viruses* (Bethesda, Md.: U.S. Department of Health and Human Services, National Institute of Allergy and Infectious Diseases, 1980), 45.

# Notes

page 114. Richard Krause's quote is from his book *The Restless Tide: The Persistent Challenge of the Microbial World* (Washington, D.C.: National Foundation for Infectious Diseases, 1981), 12.

115. Michael B. A. Oldstone's quote is from his article "Viral Alteration of Cell Function," *Scientific American*, August 1989, 42.

117. An excellent description of the mechanics of PCR was written by its inventor, Kary B. Mullis, in "The Unusual Origin of the Polymerase Chain Reaction," *Scientific American*, April 1990, 56–65.

118. The observed impairments in LCMV-infected mice occurs, Oldstone reported in *Scientific American*, because LCMV heads straight for a particular gland in a particular strain of mouse. In the LCMV-infected mice that are abnormally small, for instance, Oldstone and his colleagues found virus RNA in the very pituitary cells that ordinarily manufacture growth hormone. As a result of this impairment, the animals were producing half as much growth hormone as normal.

121. Oldstone's next step, also described in his *Scientific American* article, was to find out what it was about the virus that caused the reduction in growth hormone production. First, he traced the defect to a reduction in the cell of growth hormone messenger RNA—the genetic material that directs the manufacture of whatever protein a cell is meant to produce—rather than to a problem with the growth hormone itself. With less RNA available to orchestrate the process, less growth hormone gets made. Next, he homed in on what portion of the LCMV genome was responsible for impairing RNA transcription. "In searching for the answer, we were helped by the fact that there are two strains of LCMV, one that causes disease and one that does not," Oldstone says. Each of these viruses carries its RNA in two segments, one long and one short.

123. The work of L. Anchard and his colleagues at the Charing Cross and Westminster Medical School in London and of Heinz-Peter Schultheiss, Peter H. Hofschneider, and their colleagues at the University of Munich was described by Michael Oldstone in his *Scientific American* article, cited above. So Oldstone and his associates did a little gene swapping; their first recombinant was made of a long segment from the dangerous strain and the short segment from the safe strain; their second from a safe long segment and a dangerous short. Then they injected each hybrid LCMV into mice to see what happened. Only the viruses containing the disease-causing strain's *short* segment caused a problem. Now that the scientists have zeroed in on the region that carries the pathogenic information—which is some 3,700 nucleotides long—they can compare the safe and dangerous strains to see how the short segment differs. It is that region of difference that is most likely to contain the nucleotide sequence for pathogenicity.

124. Descriptions of the techniques for isolating the Borna virus, and its varied effects on different laboratory animals, appear in W. Ian Lipkin, Gabriel H. Travis, Kathryn M. Carbone, et al., "Isolation and Characterization of Borna Disease Agent cDNA Clones," *Proceedings of the National Academy of Sciences* 87 (June 1990): 4184–88.

# Notes

Reference to the relationship between morbid obesity and canine distemper virus in mice is from M. J. Lyons, I. M. Faust, R. B. Hemmes, et al., "A Virally Induced Obesity Syndrome in Mice," *Science* 216 (1982): 82–5.

The description of Borna disease as it affects animals is from Hanns Ludwig, Liv Bode, and George Gosztonyi, "Borna Disease: A Persistent Virus Infection of the CNS," *Progress in Medical Virology* 35 (1988): 107–51. They also used the term "apathetic" to describe some of these animals.

125. The study in the United States and Germany relating Borna disease virus antibodies to manic depression was described in R. Rott, S. Herzog, B. Fleischer, et al., "Detection of Serum Antibodies to Borna Disease Virus in Patients with Psychiatric Disorders," *Science* 228 (May 10, 1985): 755–6.

Kathryn Carbone described her study on Borna disease virus and schizophrenia in a personal interview in June 1991.

126. W. Ian Lipkin described his work with Borna disease virus in a personal interview in February 1991.

The role of inflammation in the pathogenesis of Borna disease is described in O. Narayan, S. Herzog, K. Frese, et al., "Behavioral Disease in Rats Caused by Immunopathological Responses to Persistent Borna Virus in the Brain," *Science* 220 (1983): 1401–3.

127. The theory about how an overactive immune system might cause chronic fatigue syndrome is explained in Robert Sanders, "UC San Francisco Doctors Find Way to Identify Patients with Chronic Fatigue Syndrome," *UCSF News*, November 15, 1990.

128. Anthony Komaroff's quote is from a personal interview in October 1991.

130. The CDC meeting was described in Lawrence K. Altman, "Experts Unable to Link Chronic Fatigue to Virus," *New York Times*, September 24, 1991, C5. Evidence that the syndrome is probably caused by *some* virus emerged the following month at the American Society for Microbiology annual meeting in Chicago. In a study involving ninety-two chronic fatigue patients, half of whom received an anti-virus drug for six months and half a placebo, those treated with the drug, Ampligen, recovered almost fully; those who received a placebo did not improve at all. Komaroff said the results lent weight to his conviction that chronic fatigue syndrome reflects a problem with the immune system, not the psyche. "There is no way this medicine is treating a psychological disorder," he told Associated Press reporter Daniel Q. Haney. Haney described the study, conducted by William A. Carter of the Hahnemann Medical College (a co-inventor of Ampligen), in "Anti-Virus Drug Relieves Chronic Fatigue Syndrome," *Washington Post*, October 29, 1991, Health section, 18. The relationship between HHV-6 and childhood disease is explained in "Herpes Strain Linked to High Fevers in Babies," an Associated Press report in *Washington Post*, May 28, 1992, A10.

The relationship between foamy virus and human illness is described in D. Stancek, M. Stancekova-Gressnerova, M. Janotka, et al., "Isolation and Some

Serological and Epidemiological Data on the Viruses Recovered from Patients with Subacute Thyroiditis de Quervain," *Medical Microbiology and Immunology* 161 (1975): 133–44.

page 131. Elaine DeFreitas's comments were taken from the text of a press conference held in San Francisco on September 5, 1990, in which she, Paul Cheney, and David Bell met with reporters. The transcript was reprinted in *Journal of the Chronic Fatigue and Immune Dysfunction Syndrome Association* (September 1990): 11–3. Paul Cheney's quote is also taken from this transcript. Significantly, several other investigators—including Robert Gallo of the National Cancer Institute, who discovered HTLV-II and therefore might be especially motivated to establish its connection to a common human illness—have been unable to replicate DeFreitas's findings. Gallo's work was described in Joseph Palca, "Does a Retrovirus Explain Fatigue Syndrome Puzzle?" *Science* 249 (September 14, 1990): 1240–41.

132. The work on atherosclerosis in chickens infected with Marek's disease virus is described in C. G. Fabricant, J. Fabricant, M. M. Litrenta, et al., "Virus-Induced Atherosclerosis," *Journal of Experimental Medicine* 148 (1978): 335–40. Melnick's quote is from a personal interview in September 1991.

133. In the case of mother-child transmission of herpes simplex 2 during childbirth, the baby contracts a disseminated, whole-body herpes so damaging that about half of infected newborns die. See Andre J. Nahmias, Harry Keyserling, and Francis K. Lee, "Herpes Simplex Viruses 1 and 2," in Alfred S. Evans, ed., *Viral Infections of Humans: Epidemiology and Control*, 3d ed. (New York and London: Plenum Medical Book Company, 1989), 394.

A description of the two-stage infection cycle of varicella-zoster virus can be found in Stephen E. Straus, "Clinical and Biological Differences Between Recurrent Herpes Simplex Virus and Varicella-Zoster Virus Infections," *JAMA* 262, no. 24 (December 22, 1989): 3455–8.

HHV-7 was discovered, almost by accident, in 1990 by scientists at the National Institute of Allergy and Infectious Diseases who thought they were conducting studies on the new HHV-6. By manipulating healthy blood cells to get them to divide, scientists inadvertently reactivated a latent herpesvirus infection. But when they tried to sequence the nucleic acid in the virus that appeared, they found that its genes were similar to both HHV-6 and cytomegalovirus but sufficiently different to require designation as an entirely new herpesvirus. As described in a February 2, 1990, press release from the National Institute of Allergy and Infectious Diseases ("NIAID Scientists Discover New Human Herpesvirus," by Laurie K. Doepel), the scientists were working with a blood sample from a twenty-six-year-old man who was healthy but who had had chickenpox four years before. They induced his blood cells to divide in order to get enough of them to study, to see whether varicella-zoster or any other herpesvirus was still being harbored there. Niza Frenkel, director of the infectious disease unit at the institute, noticed that when the blood cells divided, they looked odd: they formed small clumps known as syncytia, which indicated that the cell culture was harboring an infectious agent. Frenkel believed that manipulating the cells in culture is what stirred HHV-7 from its latent state—the state in which it had previously eluded all attempts to identify it. This may be the formula for the discovery of many new human viruses in the years to come, she says, especially herpesviruses, which "can sit there very quietly, not doing anything, until you start to change the host cell."

Studies conducted in recent years to support the theory that CMV is associated with atherosclerosis have been summarized in several places, most clearly in Sandra Blakeslee, "Common Virus Seen as Having Early Role in Arteries' Clogging," *New York Times*, January 29, 1991, C3. A more technical review of the literature is Joseph L. Melnick, Ervin Adam, and Michael E. DeBakey, "Possible Role of Cytomegalovirus in Atherogenesis," *JAMA* 263, no. 16 (April 25, 1990): 2204–7.

In the Stanford University study, which was described in the Melnick review, of 301 heart transplant recipients studied, 91 were infected with CMV soon after surgery. They were compared with the other 210 transplant recipients who were not infected with CMV. After five years, the survival rate was 68 percent for noninfected patients, but only 32 percent for those infected with CMV. For more details, see M. T. Grattan, C. E. Moreno-Cabral, V. A. Starnes, et al., "Cytomegalovirus Infection Is Associated with Cardiac Allograft Rejection and Atherosclerosis," *JAMA* 261 (1989): 3561–6. And for more details of the Dutch study, see M. G. R. Hendrix, M. M. M. Salimans, C. P. A. van Boven, et al., "High Prevalence of Latently Present Cytomegalovirus in Arterial Walls of Patients Suffering from Grade III Atherosclerosis," *American Journal of Pathology* 136 (1990): 23–8. The Baylor study was described in the Melnick review, as well as in B. L. Petrie, J. L. Melnick, E. Adam, et al., "Nucleic acid sequences of cytomegalovirus in cells cultured from human arterial tissue," *Journal of Infectious Diseases* 155 (1987): 158–9.

135. LeDuc's first quote is from a personal interview in January 1991. His second is from an article he wrote, "Hantaviruses Model of Emerging Agent," *U. S. Medicine*, August 1990, 41.

A description of the search for Hantavirus in urban rats comes from James LeDuc, during a personal interview, and from Richard Yanagihara, "Hantavirus Infection in the United States: Epizootiology and Epidemiology," *Reviews of Infectious Diseases* 12, no. 3 (May–June 1990): 450.

137. A description of the fieldwork in Baltimore with Hantavirus can be found in Gregory Glass, James Childs, Alan Watson, et al., "Association of Chronic Renal Disease, Hypertension and Infection with a Rat-Borne Hantavirus," *Archives of Virology*, suppl. 1 (1990): 69–80; and also in LeDuc's article in *U.S. Medicine*, cited above.

Figures about the prevalence of hypertension, and the proportion of "essential hypertension" (high blood pressure of unknown cause), are from *Merck Manual of Diagnosis and Therapy*, 15th ed. (Rahway, N.J.: Merck Sharp & Dohme Research Laboratories, 1987), 392. Figures about the annual expenditures on hypertension and related conditions are from the *Archives of Virology* article by Glass, Childs, and Watson, cited above.

## 6. Tropical Punch: The Lethal Arboviruses

page 139. The quote about the EEE victim's brain damage was made by Andrew Spielman in a personal interview in March 1991.

142. The list of exotic arboviruses was derived in large part from Robert B. Tesh, "Arthritides Caused by Mosquito-Borne Viruses," *Annual Review of Medicine* 22 (1982): 34.

The link between recreational patterns and arbovirus transmission was made in Wilbur G. Downs, "Arboviruses," in Alfred S. Evans, ed., *Viral Infections of Humans: Epidemiology and Control*, 3rd ed. (New York: Plenum Medical Book Company, 1989), 114. Andrew Spielman of Harvard points out, in a letter to the author in December 1991, that LaCrosse encephalitis viral infection is not limited to children; when adults are infected, though, they tend to be asymptomatic.

A description of the malaria outbreaks in southern California appears in Sam R. Telford III, Richard J. Pollack, and Andrew Spielman, "Emerging Vector-Borne Infections," *Infectious Disease Clinics of North America* 5, no. 1 (March 1991): 15.

144. Statistics about laboratory infection of arbovirus vectors were supplied by Thomas Monath in a personal interview, October 1991.

145. The age of the oldest recorded arbovirus infection is from William MacNeill, *Plagues and Peoples* (New York: Doubleday, 1976), 89. The notion that arboviruses have an unusually wide range, because of the movement of their vector species, was raised on page 9 of the article by Telford, Pollack, and Spielman, cited above. And information about the global impact of malaria is from two reports issued by the National Academy of Sciences: *The U.S. Capacity to Address Tropical Infectious Disease Problems* (Washington, D.C.: National Academy Press, 1987), 30; and *Malaria: Obstacles and Opportunities* (Washington, D.C.: National Academy Press, 1991).

146. Among the animal species capable of becoming sick when exposed to EEE are, curiously, some species of birds. Pheasants and whooping cranes are especially susceptible to EEE; according to Thomas Monath, one endangered population of cranes was almost wiped out in the late 1980s during a fierce epizootic of the encephalitis.

Information about the breeding preferences of the EEE bridge vectors is found on page 14 of the article by Telford, Pollack, and Spielman.

148. A description of the Tamil folk ritual of prohibiting the storage of water is in William McNeill's *Plagues and Peoples*, 235–6.

The story of the eradication of yellow fever from Cuba appears in Thomas P. Monath, "Recent Epidemics of Yellow Fever in Africa and the Risk of Future Urbanisation and Spread," *Arbovirus Research in Australia—Proceedings, 5th Symposium*, 37.

149. Monath made his comments about the living preferences of *Aedes aegypti* during a personal interview in February 1991.

The history of pesticide use in Latin America to control dengue and yellow fevers was described in Monath's article, cited above, page 37. Criticisms of such spraying programs were offered by George Craig, an entomologist at the University of Notre Dame, during a personal interview in October 1991.

The failures of programs of pest control using sterilized males was described in Rick Weiss, "The Swat Team: Gene-Altered Mosquitoes Appear on the Horizon," *Science News* 137 (February 3, 1990): 72–4. This article is also an excellent source of information regarding current programs on mosquito gene mapping.

151. George Craig criticized James Mason, former director of the U.S. Public Health Service and now under-secretary of health of the United States, for failing to stop the influx of *Aedes albopictus* when it was first observed in 1985. "He told me at the time that until it causes a human health problem, he was not interested," Craig said in an interview in October 1991. That is why he believes the U.S. practices "reactive medicine" rather than "preventive medicine."

John P. Woodall, a former director of a tropical virus laboratory founded by The Rockefeller Foundation who now works for the World Health Organization in Geneva, wrote critically of the "purists" who derided the Rockefeller labs for "stamp collecting" in a letter to the editor of *Issues in Science and Technology*, Spring 1991, 24, in response to an article on viral traffic by Stephen S. Morse.

153. The details of tracking down Oropouche virus in the 1960s were provided by Robert Shope, during a personal interview in March 1991, and appear in Max Theiler and Wilbur G. Downs, "Oropouche: The Story of a New Virus," *Yale Scientific Magazine* 38, no. 6 (March 1963).

As Robert Shope said in a personal interview in March 1991, the young Brazilian who got sick in 1961 (after volunteering to be human bait for mosquitoes in the rain forest at night) eventually recovered and went back to work—only to come down with another infection from another nighttime arbovirus.

154. A description of the cacao shell connection to Oropouche can be found in Stephen S. Morse, "Stirring Up Trouble: Environmental Disruption Can Divert Animal Viruses into People," *Sciences*, September 1990, 19.

155. The history of dengue fever comes in part from Robert Tesh of Yale during a personal interview in March 1991. Accounts from Benjamin Rush, and of the origin of the name dengue fever, appear in Donald E. Carey, "Chikungunya and Dengue: A Case of Mistaken Identity?" *Journal of the History of Medicine* 26 (1971): 243–62.

In 1989, ninety-four suspected cases of imported dengue, twenty-two of which were confirmed by blood testing, were reported to the U.S. Centers for Disease Control. See "Imported Dengue—United States, 1989," *Morbidity and Mortality Weekly Report* 39, no. 41 (October 19, 1990): 741–2.

156. Data about dengue hemorrhagic fever come from Scott B. Halstead, "Path-

*Notes*

ogenesis of Dengue: Challenges to Molecular Biology," *Science* 239 (January 29, 1988): 476. First described in Southeast Asia among children in 1956, dengue hemorrhagic fever has led to the hospitalization of at least a million and a half children, and the death of more than a quarter-million, in the years since then.

page 157. Statistics about outbreaks of dengue hemorrhagic fever in Cuba are from Halstead, cited above, page 476. Mention of the Ecuadorian and Venezuelan outbreaks of dengue hemorrhagic fever was made in John Langone, "Emerging Viruses," *Discover*, December 1990, 65. And information about the Peruvian outbreak is found in the Centers for Disease Control's *Morbidity and Mortality Weekly Reports* 40, no. 9 (March 8, 1991): 145–7.

158. Leon Rosen's quote is from "Disease Exacerbation Caused by Sequential Dengue Infections: Myth or Reality?" *Reviews of Infectious Diseases* 11, suppl. 4 (May–June 1989): S840.

Monath's quote is from a personal interview in October 1991.

According to D. Bruce Francy, an entomologist at the Centers for Disease Control's laboratory in Fort Collins, Colorado, the Asian tiger mosquito was found in Texas in 1985; in Alabama, Arkansas, Florida, Georgia, Illinois, Indiana, Louisiana, Mississippi, Missouri, Ohio, and Tennessee in 1986; and in Delaware, Kentucky, Maryland, North Carolina, and Kansas in 1987. George Craig of Notre Dame thinks it also has invaded Pennsylvania.

159. Information about 1993 mosquito abatement measures after the Midwestern floods is from Paul Cotton, "Health Threat from Mosquitoes Rises as Flood of the Century Finally Recedes," *JAMA* 270, no. 6 (August 11, 1993): 685–686.

160. The outrage at the rescheduling of the Vero Beach high school football games was voiced in Sara Rimer, "Tiny Mosquitoes' Threat Stills the Florida Night," *New York Times*, October 7, 1990, 26.

The comment by the Massachusetts state public health official was cited in Christopher B. Daly, "State's War Against Mosquitoes Leaves Some Residents Upset," *Washington Post*, September 6, 1990, A2.

### 7. The Emergence of a New Flu

163. John La Montagne's comments about influenza were made during a personal interview in March 1991. Other statistics about the great pandemic of 1918 come from Peter Radetksy, *The Invisible Invaders: The Story of the Emerging Age of Viruses* (Boston: Little, Brown, 1991), 231.

Details about the societal consequences of influenza are from Sally Squires, "Are You Ready for the Flu?" *Washington Post*, January 2, 1990, Health section, 7. The quote comparing influenza and AIDS, made by influenza researcher Peter Palese of the Mt. Sinai School of Medicine in New York City, appeared in Peter Radetsky's book, cited above, page 231.

164. Edwin D. Kilbourne made his comments during a personal interview in March 1991.

## Notes

The symptoms of ordinary and pandemic influenza appear in *The Merck Manual of Diagnosis and Therapy* 15th ed. (Rahway, N.J.: Merck, Sharp & Dohme Research Laboratories, 1987), 172.

166. Statistics on the number of American deaths from the Asian and Hong Kong flu are in Diana B. Dutton, *Worse Than the Disease: Pitfalls of Medical Progress* (Cambridge: Cambridge University Press, 1988), 133.

167. Holland based his calculations on work with the vesicular stomatitis virus—an animal virus that, like influenza, is based on RNA rather than DNA. His experiments are described in Kathleen McAuliffe, "The Killing Fields," *Omni*, 1990, 94, and also in Julie Anne Miller, "Diseases for Our Future," *BioScience* 39, no. 8 (September 1989): 512.

Other human viruses also are made of segmented genomes, among them rotaviruses, which cause diarrhea; arenaviruses, which cause Lassa fever, Argentinian hemorrhagic fever, and Bolivian hemorrhagic fever; and Hantaviruses. They are capable of reassorting in much the same way as influenza, but because they are more restricted in their geographic range—and less likely to encounter animal species capable of harboring more than one viral strain at a time—this method of recombination is less of a threat to human health. For more information, see Edwin D. Kilbourne, "New Viral Diseases: A Real and Potential Problem Without Boundaries," *JAMA* 264, no. 1 (July 4, 1990): 68–70.

168. The description of pigs as "mixing vessels" is from Julie Ann Miller's article, cited above, page 513. The leading proponents of this theory are Robert G. Webster of the St. Jude's Children's Research Hospital in Memphis and Christoph Scholtissek of the Institut für Virologie at Justus-Liebig-Universität in Giessen, Germany.

169. Stephen Morse made his comments about pandemic influenza in his article "Regulating Viral Traffic," in *Issues in Science and Technology*, Fall 1990, 82.

170. The European scientists' concern about fish farming was expressed in a letter to the editor of *Nature* 331 (1989): 215.

172. Kilbourne's op-ed article, from which this quote was taken, was headlined, "Flu to the Starboard! Man the Harpoons! Fill 'em with Vaccine! Get the Captain! Hurry!" It appeared, along with a large cartoon depicting the crew of an old sailing vessel surrounded by raging seas and the menacing long arms of disease, in *New York Times*, February 13, 1976, 33.

173. The first paper about the genetically "frozen" Russian flu virus, which was discussed by Brian Murphy in a personal interview in November 1991, was written by Peter Palese and his colleagues at Mt. Sinai. His work is summarized in Peter Radetsky, *The Invisible Invaders* (Boston: Little, Brown, 1991), 241.

174. Fred Hoyle and N. Chandra Wickramasinghe explain their theory about sunspot activity and pandemic influenza in a letter to the editor, "Sunspots and Influenza," *Nature*, January 25, 1990.

The theory that hemagglutinin types cycle in sequence was offered by Brian Murphy during a personal interview in November 1991. Information about antibodies in people born in the late nineteenth century, and studied in the mid-1970s, can be found in W. Paul Glezen and Robert B. Couch, "Influenza Viruses," in Alfred S. Evans, ed., *Viral Infections of Humans: Epidemiology and Control*, 3rd ed. (New York and London: Plenum Medical Book Company, 1989), 432.

page 175. The history of Reye's syndrome, including Douglas Reye's early observations, is described in Evelyn Zamula, "Reye Syndrome: The Decline of a Disease," *FDA Consumer* (November 1990): 21–3. Lawrence Schonberger's quote is from this article as well.

178. Kilbourne made his comment about the swine flu immunization program in a letter to the author in January 1992.

179. For an excellent summary of both the scientific and political issues raised by the "swine flu fiasco," see the chapter on swine flu in Diana B. Dutton, *Worse Than the Disease*, cited above, 127–73. Many of the details presented here are derived from Dutton's chapter, which she wrote with the assistance of J. Bradley O'Connell and Thom Seymour.

Edwin D. Kilbourne reviewed the lessons of his experience with the swine flu immunization campaign in an article called "Swine Flu: The Virus That Vanished," *Human Nature*, March 1979, 73.

A description of the CDC expedition to China in 1987 appears in Peter Jaret, "The Disease Detectives," *National Geographic*, January 1991, 116–40.

180. Kilbourne made his comments about surveillance during a personal interview in March 1991 and also in a commentary in response to Stephen Morse's article on viral traffic, published in *Issues in Science and Technology*, Spring 1991, 22–3.

## 8. Anticipating the Next AIDS

183. Information about the sentinel chickens in Florida comes from the report by experts at the Centers for Disease Control, "Arboviral Surveillance—United States, 1990," in *Morbidity and Mortality Weekly Report* 39, no. 35 (September 7, 1990): 593–7.

184. The 1993 spraying in central Florida was mentioned in a column called "Around the Nation" in *The Washington Post*, July 1, 1993, A4.

D. A. Henderson's proposal for a global surveillance system is described in Stephen Morse's article "Regulating Viral Traffic," in *Issues in Science and Technology*, Fall 1990, 83. Estimates for the cost of care of AIDS patients appear in Malcolm Gladwell, "U.S. Visa Policy Denounced at Global AIDS Conference," *Washington Post*, June 20, 1991, A3.

185. D. A. Henderson expressed his ideas about tropical disease training centers in comments to Mitchel L. Zoler of *Medical World News*, who cited him in his article "Old Diseases Are New to Many Doctors," June 25, 1990, 23. He also contributed to the report from the National Academy of Sciences that reached a similar conclusion: *The U.S Capacity to Address Tropical Infectious Disease Problems* (Washington, D.C.: National Academy Press, 1987).

186. Wilbur G. Downs's comments come from his chapter, "Arboviruses," in Alfred S. Evans, ed., *Viral Infections of Humans: Epidemiology and Control*, 3rd ed. (New York: Plenum Medical Book Company, 1989), 109.

187. Robert Shope made his comments during a personal interview in March 1991.

Jonathan Mann's comments were made during a personal interview in April 1991.

188. John P. Woodall's quote is from his letter to the editor of *Issues in Science and Technology* 7, no. 3 (Spring 1991): 24.

Stephen Morse's quote is from his article "Regulating Viral Traffic," *Issues in Science and Technology*, Fall 1990, 84.

Joshua Lederberg's quote is from his article "Pandemic as a Natural Evolutionary Phenomenon," *Social Research* 55, no. 3 (Autumn 1988): 358.

190. The story of Wil Downs's involvement in the Senegal River dam project was told by Robert Shope during a personal interview in March 1991. Details about the outbreak in Mauritania are from John Walsh, "Rift Valley Fever Rears Its Head," *Science* 240 (June 10, 1988): 1397–9.

192. Karl Johnson tells the story of the appearance and disappearance of Bolivian hemorrhagic fever in his chapter, "Emerging Viruses in Context: An Overview of Viral Hemorrhagic Fevers," in Stephen Morse's book *Emerging Viruses* (New York: Oxford University Press, 1992). He considers it in part a cautionary tale, saying, "Let this be a lesson for all who desire to build a research program on a single new disease!" His point is that the disease might easily disappear long before the research program gets under way.

193. Edwin Kilbourne made his comments during a personal interview in March 1991. In addition, he wrote some of his thoughts in a letter to the editor in response to Stephen Morse's article in *Issues in Science and Technology*. Kilbourne's response, as well as those of several other scientists, appeared in that journal's Spring 1991 issue (vol. 7, no. 3), on page 22.

194. Details about the experience monitoring influenza in collaboration with Chinese scientists are from Teri Randall, "CDC Plays Cat-and-Mouse With Flu Virus," *JAMA* 263, no. 19 (May 16, 1990): 2574–9.

195. Jonathan Mann made his comments about hubris during a personal interview in April 1991.

Jonathan Mann made his comments about creative surveillance strategies in his article "AIDS and the Next Pandemic," *Scientific American*, March 1991, 126. He also expanded upon these ideas in a personal interview in April 1991.

196. Information about the Biological Weapons Convention of 1972, and other details about biological warfare, come from Melissa Hendricks, "Germ Wars," *Science News* 134 (December 17, 1988): 392–5. The quote about when offensive biological weapons research is permissible is from a 1969 statement issued by Henry Kissinger, then the U.S. national security adviser, and cited in that article.

page 197. Mark Wheelis made his comments about biological warfare in one of the letters to the editor written in response to Stephen Morse's article. His letter was published in *Issues in Science and Technology* 7, no. 3 (Spring 1991): 21.

## 9. Viral Domestication

198. Details about W. French Anderson and his patients, as well as all his quotes, were gathered during interviews conducted for the article "Dr. Anderson's Gene Machine," by Robin Marantz Henig, *New York Times Magazine*, March 31, 1991, 31–5 ff.

200. René Dubos's quotes are from *The Unseen World* (New York, The Rockefeller Institute Press, 1962), 71, 89.

201. Richard Mulligan's quote is from Peter Radestsky, *The Invisible Invaders: The Story of the Emerging Age of Viruses* (Boston: Little, Brown, 1991), 378. Anderson is careful to correct anyone who calls the process of "transduction" by its more familiar name, "infection," even though it basically describes what happens in the culture dish. He says the word infection is "too emotive."

203. Details about David's final days are from Boyce Rensberger, "Viruses: The Lifeless Invaders That Enslave Cells and Kill Us," *Washington Post*, August 18, 1985, B1.

204. Ronald Crystal's quote, and the details of his experiment, are from Michelle Hoffman, "New Vector Delivers Genes to Lung Cells," *Science* 252 (April 19, 1991): 374. His work was described on pages 431–4 in the same issue of *Science*, in an article he co-authored with senior authors Melissa A. Rosenfeld, Wolfgang Siegfried, et al., "Adenovirus-Mediated Transfer of a Recombinant Alpha 1-Antitripsin Gene to the Lung Epithelium in Vivo."

205. Barrie Carter made his comment at a public meeting of the NIH Recombinant DNA Advisory Committee, which met in Bethesda, Maryland, in November 1990.

207. A description of intracellular immunization, using both VP-16 genes and *tat* genes, can be found in Rick Weiss, "Well-Bred Cells: Poor Hosts to Viruses," *Science News* 134 (October 1, 1988): 213.

209. Abram Benenson's quote, as well as details of the last case of smallpox on earth, are from his chapter, "Smallpox," in Alfred S. Evans, ed., *Viral Infections of Humans: Epidemiology and Control*, 3rd ed. (New York: Plenum Medical Book Publishing, 1989), 633.

The story of Edward Jenner's early experimentation with smallpox vaccine comes primarily from Peter Radetsky's book *The Invisible Invaders*, cited above.

210. Frank Fenner made his comments about the origin of vaccinia virus in the foreword to the book by Derrick Baxby, *The Origin of Vaccinia Virus*, page vii.

211. Information about the vaccinia-based rabies vaccine being used in France and Belgium is from Malcolm W. Browne, "New Animal Vaccines Spread Like

Diseases," *New York Times*, November 26, 1991, C1. A report of the follow-up study in Belgium appeared in Boyce Rensberger, "Vaccinating Foxes to Protect Cows," *Washington Post*, December 23, 1991, A2. Although the recombinant vaccinia rabies vaccine seems effective in foxes, the *Post* article pointed out that it does not work in raccoons or skunks, the two chief carriers of rabies in the eastern United States.

213. Joel Dalrymple's quote about vaccinia virus vectors is from a personal interview in March 1991.

A description of the side effects that plagued the global eradication campaign for smallpox can be found in Abram Benenson's chapter, "Smallpox," cited above, on page 650. "These complications assumed real importance where smallpox did not exist, so that it could truly be said that the preventive measure was costing more lives than the disease itself," Benenson wrote. "Unfortunately, emphasis in the Western world on these complications resulted in fear of vaccination and over-response when the unimportant reactions were seen."

214. The controversial AIDS vaccine research, conducted by Daniel Zagury of the Pierre and Marie Curie Institute in Paris, was summarized in Alexander Dorozynski and Alun Anderson, "Deaths in Vaccine Trials Trigger French Inquiry," *Science* 252 (April 16, 1991): 501–2.

Bernard Moss's work with vaccinia viruses is described in a press release written by Laurie K. Doepel and issued by the National Institute of Allergy and Infectious Diseases on November 20, 1990.

215. Information about the Australian rabbit/fox eradication program and the quote from Mark Bradley are from Virginia Morell, "Australian Pest Control by Virus Causes Concern," *Science* 261 (August 6, 1993): 683–684.

## 10. Toward a New Biology

217. The description of the "medical war game" played by Llewellyn Legters comes from a chapter he wrote with Linda Brink and Ernest Takafuji, "Are We Prepared for a Viral Epidemic Emergency?" in Stephen S. Morse, ed., *Emerging Viruses* (New York: Oxford University Press, 1992).

219. According to Stephen Morse, Hans Stetten drew his hard-soft line during a lecture he gave at the Marine Biological Laboratory in Woods Hole, Massachusetts. Morse was not at the lecture, but the story of the line has become apocryphal among biomedical scientists.

220. Walter Gilbert's quotes are from his commentary, "Towards a Paradigm Shift in Biology," *Nature* 349 (January 10, 1991): 99.

221. Carleton Gajdusek made his remarks about molecular biology during a personal interview in December 1991.

222. Stephen Morse's comment about retroviruses as laboratory tools was made during a personal interview in March 1991.

223. The story of polio typing was told by Gajdusek during the December 1991 interview.

page 223. The quote from Keith Stewart Thomson is from his article "Reductionism and Other Isms in Biology," *American Scientist* 72 (1984): 388–90.

224. Bernard Fields made his comments about the activation of reovirus during a personal interview in March 1991.

226. The story of Walter Reed's adventures in Cuba appears in Carol Eron, *The Virus That Ate Cannibals: Six Great Medical Detective Stories* (New York: Macmillan Publishing, 1981), 18–19. Among the members of the Reed Commission who died after volunteering to be bitten by disease-carrying mosquitoes were Jesse Lazear, an entomologist, and Clara Maass, a nurse.

227. Statistics about medical student debt are from Janet Bamford, "Doctors of Finance," *Forbes* 137 (April 7, 1986): 123.

228. The National Academy of Sciences report on the future of biology was the subject of an article by Stu Borman, "Biology Research: Report Stresses Interdisciplinary Ties," in *Chemical and Engineering News* (December 18, 1989): 4.

229. Stephen Morse discussed the pros and cons of a new field of "traffic science" during the March 1991 interview.

230. Lewis Thomas's quote about Woods Hole is from *The Lives of a Cell: Notes of a Biology Watcher* (New York: Bantam Books, 1974), 73.

Mathilde Krim's comments about AIDS were made during a speech she gave to the graduating class of the State University of New York at Stony Brook, which awarded her an honorary degree on May 19, 1991. Excerpts of the speech appeared in *The Scientist* (June 24, 1991): 11.

# Bibliography

Balows, Albert, editor-in-chief. *Manual of Clinical Microbiology*, 5th ed. Washington, D.C.: American Society for Microbiology, 1991.

Burnet, Sir Macfarlane. *Natural History of Infectious Disease*, 3rd ed. New York and London: Cambridge University Press, 1962.

Dawkins, Richard. *The Selfish Gene*. Oxford and New York: Oxford University Press, 1989.

Dubos, René. *Mirage of Health: Utopias, Progress, and Biological Change*. New York: Harper & Row, 1959.

Dubos, René and Jean Dubos. *The White Plague: Tuberculosis, Man and Society*. Boston: Little, Brown, 1952.

Duncan, Ronald, and Miranda Weston-Smith, eds. *The Encyclopedia of Medical Ignorance*. Oxford: Pergamon Press, 1984.

Dutton, Diana B. *Worse Than the Disease: Pitfalls of Medical Progress*. Cambridge: Cambridge University Press, 1988.

Dwyer, John M. *The Body at War: The Miracle of the Immune System*. New York: New American Library, 1988.

Eron, Carol. *The Virus That Ate Cannibals: Six Great Medical Detective Stories*. New York: Macmillan Publishing, 1981.

Evans, Alfred S., ed. *Viral Infections of Humans: Epidemiology and Control*. New York and London: Plenum Medical Book Company, 1989.

Fee, Elizabeth, and Daniel M. Fox, eds. *AIDS: The Burdens of History*. Berkeley and Los Angeles: University of California Press, 1988.

Fenner, Frank, B. R. McAuslan, C. A. Mims, et al. *The Biology of Animal Viruses*, 2nd ed. New York and London: Academic Press, 1974.

Gleick, James. *Chaos: Making a New Science*. New York: Penguin Books, 1987.

Grmek, Mirko D. *History of AIDS: Emergence and Origin of a Modern Pandemic*. Trans-

## Bibliography

lated by Russell C. Maulitz and Jacalyn Duffin. Princeton, N.J.: Princeton University Press, 1990.

Hoyle, Fred, and Chandra Wickramasinghe. *Diseases from Space*. New York: Harper & Row, 1979.

Krause, Richard M. *The Restless Tide: The Persistent Challenge of the Microbial World*. Washington, D.C.: The National Foundation for Infectious Diseases, 1981.

Lederberg, Joshua, Robert E. Shope, and Stanley C. Oaks, Jr., eds. *Emerging Infections: Microbial Threats to Health in the United States*. Washington, D.C.: National Academy Press, 1992.

McNeill, William H. *Plagues and Peoples* New York: Anchor Press/Doubleday, 1976.

Morse, Stephen S. *Emerging Viruses*. New York: Oxford University Press, 1992.

National Academy of Sciences. *The U.S Capacity to Address Tropical Infectious Disease Problems*. Washington, D.C.: National Academy Press, 1987.

Radetsky, Peter. *The Invisible Invaders: The Story of the Emerging Age of Viruses*. Boston: Little, Brown, 1991.

Scott, Andrew. *Pirates of the Cell: The Story of Viruses from Molecule to Microbe*. New York: Basil Blackwell, 1987.

Shilts, Randy. *And the Band Played On: Politics, People, and the AIDS Epidemic* (revised edition). New York: Penguin Books, 1988.

Shoumatoff, Alex. *African Madness*. New York: Alfred A. Knopf, 1988.

Silverstein, Arthur M. *A History of Immunology*. San Diego, Calif.: Academic Press, Inc., 1989.

Thomas, Lewis. *Late Night Thoughts on Listening to Mahler's Ninth Symphony*. New York: Bantam Books, 1984.

Thomas, Lewis. *The Lives of a Cell: Notes of a Biology Watcher* New York: Bantam Books, 1974.

Wills, Christopher. *The Wisdom of the Genes: New Pathways in Evolution*. New York: Basic Books, 1989.

Witt, Steven C. *Biotechnology, Microbes and the Environment* San Francisco, Calif.: Center for Science Information, 1990.

# Index

# Index

# Index

Montagnier, Luc, 47
Montali, Richard J., 102
morbillivirus, 88, 89–94, 114
Morse, Stephen, xiii–xiv, 12–19, 33–6, 39–40, 51, 87–8, 168–9, 175, 184–9, 222, 228–9
mosquitoes, *Aedes aegypti*, 11, 12, 147–9, 151, 155, 158; *Aedes albopictus* (tiger mosquito), 12, 158–9; *Aedes bendersoni*, 150; *Aedes triseriatus*, 150; *Aedes vexans*, 146–7; *Coquillettidia perturbans*, 146–7; *Culex*, 25–6, 142–3, 153, 159; *Culiseta melanura*, 145–6; *see also* arboviruses
Moss, Bernard, 214–15
Mulligan, Richard, 201
multiple sclerosis, 29
Murphy, Brian, 173
myosistis, 177
myxomatosis virus, 6–9, 18, 44, 88

Nathanson, Neal, 103
National Academy of Sciences, xv, 39, 188
National Cancer Institute, 104, 198
National Institute of Allergy and Infectious Diseases (NIAID), xiii, 16, 55, 162, 173, 214
National Institutes of Health (NIH), xiii, 18, 40, 47, 198–207, 218, 221
natural killer cell, 71, 118
Navajo, viral outbreak, x
neuraminidase, 170–5
*New York Times, The*, 16, 172–3
Nobel Prize, 13, 18
nonspecific immune responses, 71
Norway, viruses in, 51–2, 53
Norwood, William, 54–5
nucleic acid hybridization, 116–17
nucleotides, 65–8, 96, 115–19, 156, 166–7, 211, 221

Oldstone, Michael, 115–18, 121, 122
*Omni* (magazine), 17
organ transplants, 28–31, 122, 133, 206; and AIDS transmission, 54–5
Oropouche virus, 141, 142–4

pandemic influenza, 162–80
Papua New Guinea, xv
Parrish, Colin, 96–9
parvoviruses, 62, 88, 94–9
Pasteur, Louis, 61
pathogens, basic principles of, 59–81
PEG-ADA therapy, 203–5
peste des petits ruminants virus, 90, 92

pesticides, 148–9, 159, 160
Philippines, 85–7
picornavirus, 90
pituitary gland, 117–18, 121, 123
plant viruses, 59–60
Platt, Sir Robert, 50
*Pneumocystis* pneumonia, 37, 38, 49–50
point mutations and deletions, 166–7
poliovirus, xii, 32, 39, 46, 63, 69, 105, 107, 113, 130, 165, 223
polymerase chain reaction, 51, 55, 117–19
Potosí virus, 159
poxviruses, 208–15
PR-8 influenza, 177
primary viral pneumonia, 164
prostitution, and AIDS, 42, 46, 48, 49, 53
proteins, 64–70, 72, 94–5, 113, 117–18, 221; viral, 59, 62–3, 64, 69–70, 94–5, 105, 113–14, 165, 166, 167, 200, 207–15
proteinuria, 11

rabbit virus, 6–9, 44, 88
rabies virus, 69, 75, 211, 213
Rask, Margrethe, 37, 38
recombinant vaccinia vaccine, 211–15
reductionism, in molecular biology, 220–4
Reed, Walter, 226
reovirus, 69, 224
replication, 65–8, 80, 95, 167, 200
retroviruses, 64, 69–71, 77–80, 104–5, 130–1, 199–204, 213, 222; endogenous, 80, 104–5; HIV, 37–57, 80–1, 104–5, 116–17, 194, 214–15
Reye, Douglas, 175–6
Reye's syndrome, 175–7
rhinoviruses, 62, 75, 204
ribavirin, 5–6
Rift Valley fever, 28, 190–1
rinderpest virus, 90, 91
RNA (ribonucleic acid), 59, 63–70, 115–16, 123, 156, 166–7, 214–15, 222; viruses, 63–4, 77–80, 115–16, 166–7
Rocio virus, 141
Rockefeller Foundation, New York, 91, 153, 188, 193
Rockefeller University, New York, xv, 12–14, 88, 152
rodent-borne diseases, 4–14, 20–4, 27–9, 56–7, 64, 76, 101–5, 108–9, 118–27, 130, 135–8, 142, 154, 160, 189–94, 199, 201, 205, 207
Rosen, Leon, 157–8
Rosenberg, Steven, 199
roseola, 130, 131
Ross River virus, xv
Rous sarcoma virus, 70